Ergebnisse der Mathematik und ihrer Grenzgebiete

3. Folge · Band 5

A Series of Modern Surveys in Mathematics

Kurt Strebel

Quadratic Differentials

With 74 Figures

Springer-Verlag
Berlin Heidelberg New York Tokyo 1984

Kurt Strebel
Mathematisches Institut
Universität Zürich
Rämistraße 74
CH-8001 Zürich

AMS Subject Classification (1980): 30-01,
30Cxx, 30C70, 30C75, 30Fxx

ISBN 3-540-13035-7 Springer-Verlag Berlin Heidelberg New York Tokyo
ISBN 0-387-13035-7 Springer-Verlag New York Heidelberg Berlin Tokyo

Printed in Germany

Typesetting, printing and bookbinding:
Universitätsdruckerei H. Stürtz AG, Würzburg.
2141/3140-543210

Preface

A quadratic differential on a Riemann surface is locally represented by a holomorphic function element which transforms like the square of a derivative under a conformal change of the parameter. More generally, one also allows for meromorphic function elements; however, in many considerations it is convenient to puncture the surface at the poles of the differential. One is then back at the holomorphic case.

A quadratic differential defines, in a natural way, a field of line elements on the surface, with singularities at the critical points, i.e. the zeros and poles of the differential. The integral curves of this field are called the trajectories of the differential.

A large part of this book is about the trajectory structure of quadratic differentials. There are of course local and global aspects to this structure. Besides, there is the behaviour of an individual trajectory and the structure determined by entire subfamilies of trajectories.

An Abelian or first order differential has an integral or primitive function on the surface, which is in general not single-valued. In the case of a quadratic differential, one first has to take the square root and then integrate. The local integrals are only determined up to their sign and arbitrary additive constants. However, it is this multivalued function which plays an important role in the theory; the trajectories are the images of the horizontals by single valued branches of its inverse.

A quadratic differential also defines, in a natural way, an element of arc length on the surface, and a corresponding area element. The associated metric and its geodesics are studied in detail in this volume. The trajectories are "horizontal" geodesics. The total area of the surface with respect to this metric is nothing but the L^1-norm of the differential.

We start with a chapter about Riemann surfaces. The main object is the annular covering surface associated with a closed curve, in particular a Jordan curve, on the underlying surface. This covering surface is an important tool in connection with closed trajectories and the convergence of ring domains. But it is only needed in the later chapters and the reader who is familiar with Riemann surfaces will of course omit that part of the book.

The local theory of trajectories and of the associated metric makes use of special parameters, in terms of which the differential has a particularly simple representation.

The integration of the field of line elements is performed by means of the above mentioned integral function or rather its inverse. The theory of differential equations on the Riemann surface is thus replaced by conformal mapping and analytic continuation. The most important advantage of this procedure lies in the fact, that the integration of one trajectory is extended, automatically, to the neighboring ones.

The most important result about geodesics is the fact that they are the unique shortest curves in their homotopy class, with the same endpoints. Closed trajectories are shortest in their free homotopy class. This is treated, together with supplementary results, in the Chapter about the metric. In connection with the length-area method it is natural to postulate finite area, i.e. finite norm of the differential; but this is not so if one merely considers geodesics.

A special Chapter is devoted to quadratic differentials the trajectories of which are closed, except for a set of measure zero. There are three types of problems, according to the data (besides the topological ones): the lengths of the circumferences of the cylinders, their heights, or the ratio of their moduli. The crucial distinction is between finite and infinite curve systems. In the first case, all three problems are in a certain sense equivalent, and they are treated in full detail. In the infinite case, the first two problems (and the problem of the extremal metric closely connected with problem one) are treated. This is possible because the solutions are given by quadratic differentials of finite norm, whereas the problem of moduli leads outside if this set. Only very little is known about this problem, and a restriction to differentials of finite norm would seem quite unnatural.

The last Chapter of the book deals with quadratic differentials of general type and with their approximation by quadratic differentials with closed trajectories. The main result of this part is a uniqueness theorem, the so-called heights theorem.

It has turned out that quadratic differentials and their trajectories are connected with a large variety of function theoretic extremal problems. They first appeared in M. Schiffer's variational calculus for schlicht functions as solutions of differential equations for the boundary curves. But the real impact came from O. Teichmüller's work on extremal quasiconformal mappings in his search for some general principle underlying H. Grötzsch's examples of mappings with smallest maximal dilatation. His beautiful work ultimately led to a solution of the moduli problem for Riemann surfaces and to the Teichmüller spaces. Quite recently, the trajectories of quadratic differentials found their place in the theory of measured foliations.

There exist few presentations of the theory in book form, and these usually serve a specific purpose and are tailored accordingly. Most of the basic ideas can be found, often with only a sketch of a proof, in Teichmüller's work: the greater part occurs in [1], [2] and [3] of our list of references. Schaeffer and Spencer's work [1] treats schlicht functions and therefore deals with rational quadratic differentials in the plane. Jenkins [3] presents a theory of meromorphic quadratic differentials on finite Riemann surfaces which are real along the boundary. Then, there are several fragments of the theory in lecture notes (Ahlfors [1], Pfluger [1], Strebel [7]) and in short chapters of books (Ahlfors [3], Jensen [1]). All these presentations start with a description of the local trajectory structure and proceed to some theorems about geodesics and trajectories in the large. Besides these, further articles by various authors are scattered in the mathematical literature.

In the light of this situation it seemed to be desirable to collect and round up the basic facts of the theory in a book. This is an attempt in this direction. The presentation is sometimes quite broad, and I have not hesitated to give more than one proof of the same theorem, if the methods are very different and shed light on a fact from alternative sides.

On the other hand, various topics have come short. Quadratic differentials entering into the theory of schlicht mappings have not been treated specially. I refer to Teichmüller [2], to the book by Schaeffer and Spencer [1], to chapter 8 of Ahlfors [3] or to Jensen's account in Pommerenke [1]. Anything connected to Jenkins' general coefficient theorem can of course be looked up in his book. The topological extensions of the theory of trajectories are omitted, as are the recent developments of the theory related to interval exchange transformations and ergodicity (Masur [2] and [3]). Our basic attitude in the book is classical in the sense that quadratic differentials are considered on a fixed Riemann surface: thus Teichmüller theory is another topic which has not even been touched.

The book grew out of my Lecture Notes: On quadratic differentials and extremal quasiconformal mappings (Minneapolis 1967). I am grateful to the University of Minnesota and to Stanford University for the opportunities to lecture on the subject in the years 1967 and 1973 respectively, as well as to the University of Maryland where I stayed in the fall of 1981. The joint work with E. Reich and A. Marden has greatly stimulated my interest in the field; the last chapter of the book presents results of a joint paper with A. Marden.

I wish to express my gratitude to the editors for accepting the account in this renowned series as well as to the staff of Springer for the efficient handling of the editing.

Finally, I am indebted to Dr. Madeleine Sitterding for the great care she took in typing the manuscript.

November 1983

K. Strebel
University of Zürich

Table of Contents

Chapter I. Background Material on Riemann Surfaces

§ 1. Riemann Surfaces

1.1 Definition of a Riemann surface. In this paragraph some basic facts about Riemann surfaces are recalled. For the general theory we refer to Ahlfors and Sario [1], Nevanlinna [1], Lehto [1] and Strebel [15].

Definition 1.1. A Riemann surface R is a connected Hausdorff space M together with an open covering $\{U_\nu\}$ and a system of homeomorphisms h_ν of the sets U_ν onto open sets $V_\nu = h_\nu(U_\nu)$ in the complex plane $\mathbb{C}$ with conformal neighbor relations. The latter means: whenever $U_\nu \cap U_\mu \neq \emptyset$, the composition $h_\nu \circ h_\mu^{-1}$ is a direct conformal homeomorphism of the open set $h_\mu(U_\nu \cap U_\mu)$ onto $h_\nu(U_\nu \cap U_\mu)$.

We will sometimes write $R = (M, \{(U_\nu, h_\nu)\})$ for this Riemann surface.

The space M is a two dimensional manifold, because every point $P \in M$ has a neighborhood which is homeomorphic to an open set in the plane. The system $\{(U_\nu, h_\nu)\}$ is called a conformal structure on the manifold M. An arbitrary pair (U, h), where U is an open set in M and h is a homeomorphism of U onto an open set V in the plane, is said to be compatible with the conformal structure $\{(U_\nu, h_\nu)\}$, if the neighbor relation of h with h_ν is conformal whenever $U \cap U_\nu \neq \emptyset$. In other words when we can join the pair (U, h) to the system (U_ν, h_ν) and still have a conformal structure. The homeomorphism h: $U \to V$ is called a uniformizer in U or a local uniformizer, and the complex variable $z \in V$ is called a (local) parameter in U or near P for any $P \in U$.

Two conformal structures $\{(U_\nu, h_\nu)\}$ and $\{(\tilde{U}_\mu, \tilde{h}_\mu)\}$ on the same manifold M are called equivalent, if their union still is a conformal structure. This means of course that every pair (U_ν, h_ν) is compatible with the structure $\{(\tilde{U}_\mu, \tilde{h}_\mu)\}$ and vice versa. Two Riemann surfaces $R = (M, \{(U_\nu, h_\nu)\})$ and $\tilde{R} = (M, \{(\tilde{U}_\mu, \tilde{h}_\mu)\})$ with the same underlying manifold M and equivalent conformal structures are called equal.

If we start out with a given conformal structure $\{(U_\nu, h_\nu)\}$ on M we can add all possible pairs (U, h) which are compatible with the given structure. This completed system is equivalent to the given one. Two Riemann surfaces are equal if and only if the underlying manifolds are the same and the maximal (completed) conformal structures are identical.

1.2 Bordered Riemann surfaces

Definition 1.2. A bordered Riemann surface R is a connected Hausdorff space M together with an open covering $\{U_\nu\}$ of M and corresponding homeomorphisms h_ν, h_ν: $U_\nu \to V_\nu = h_\nu(U_\nu)$, where V_ν is a relatively open set of a closed half plane and where the neighbor relations again are conformal.

We can assume this half plane to be the set $y_\nu \geq 0$ in the complex z_ν-plane, $z_\nu = x_\nu + i\,y_\nu$. If a point $P \in M$ is mapped onto a point x_ν on the real axis by a homeomorphism h_ν, this is the case for every homeomorphism h_μ defined at P. The set of these points P is said to form the border Γ of the Riemann surface R, and we assume of course that it is not empty on a bordered Riemann surface. The border Γ is a set of points on R, not to be confused with the ideal boundary of R. (A sequence of points P_n of R tends to its ideal boundary if and only if it has no accumulation point in R.) The set Γ, together with the corresponding restrictions of the homeomorphisms h_ν, is a – not necessarily connected – one dimensional manifold. The conformality of the neighbor relation here means that at the interior points of V_ν, i.e. points which are not on the real axis, the composed mapping $h_\mu \circ h_\nu^{-1}$ is conformal, if defined.

1.3 The Mirror image of a Riemann surface, the double of a bordered Riemann surface. Every Riemann surface R, whether bordered or not, has a mirror image (symmetric image). Let $R = (M, \{(U_\nu, h_\nu)\})$. Then the mirror image of R is defined to be the surface

$$R^* = (M, \{(U_\nu, \bar{h}_\nu)\}),$$

with $\bar{h} = s \circ h_\nu$, s the reflection of the complex plane on the real axis. R^* is a Riemann surface, because the compositions $\bar{h}_\nu \circ \bar{h}_\mu^{-1}$ are orientation preserving conformal homeomorphisms whenever they have a meaning.

It is often preferable to think of two replicas of the underlying manifold, say M and M^*. The identity mapping $P \leftrightarrow P^*$ is called the symmetry of R onto R^*. It is an anti-conformal homeomorphism. The mirror image R^* only depends on R and not on the particular system $\{(U_\nu, h_\nu)\}$ defining its conformal structure.

The mirror image R^* of a bordered surface can be glued along some open intervals Γ_λ of the border Γ to the original surface R to form a new surface $\tilde{R}$. This surface is called the double of R along $\{\Gamma_\lambda\}$; if $\{\Gamma_\lambda\} = \Gamma$ we just speak of the double of R. The set $\tilde{M}$ consists of the union of M and M^* with the points $P \in \Gamma_\lambda$ identified with their symmetric images P^* on M^*. The open covering of $\tilde{M}$ consists of all the open sets U_ν and their symmetric images U_ν^*, again with the identification $P_\nu = P_\nu^*$ if $P_\nu \in U_\nu$ is a point of some Γ_λ. The two homeomorphisms h_ν of U_ν and h_ν^* of U_ν^* agree on the border, in particular on the set Γ_λ, and because of the Cauchy integral formula the interval on the real line is "removable". Hence the neighbor relation is conformal throughout the glued neighborhood. The double of R along $\{\Gamma_\lambda\}$ is simply the double of R, cut along the set $\Gamma \setminus \{\Gamma_\lambda\}$.

1.4 Holomorphic mappings. A holomorphic (analytic) mapping f of a Riemann surface R into a Riemann surface R' is a continuous mapping, the representation of which in terms of local parameters near P and its image $P'=f(P)$, is a holomorphic function.

The notion of direct analytic continuation is the same as it is for analytic functions: let G_1 and G_2 be two domains of a Riemann surface R with non void intersection. A holomorphic mapping f_2 of G_2 into a Riemann surface R' is called direct analytic continuation of the holomorphic mapping f_1: $G_1 \to R'$, if $f_2 = f_1$ on $G_1 \cap G_2$. The uniqueness of a direct analytic continuation is proved as in function theory.

§2. Jordan Curves on Riemann Surfaces

2.1 The fundamental group, covering surfaces. Let R be an arbitrary Riemann surface, $P_0 \in R$. A closed path γ with initial point P_0 is a continuous mapping of the unit interval $0 \le t \le 1$ into R, $P = \gamma(t)$, where $\gamma(0) = \gamma(1) = P_0$. The homotopy class of γ will be denoted by $c = [\gamma]$. The fundamental group $F = F_{P_0}$ of R with base point P_0 is the group of all homotopy classes of closed paths with initial point P_0, with the usual multiplication by composing the closed paths.

The fundamental groups F_{P_0} and F_{P_1} with base points P_0 and P_1 respectively are isomorphic. A canonical isomorphism is induced by any path α joining P_0 to P_1 and assigning the path $\gamma_0 = \alpha \gamma_1 \alpha^{-1}$ with initial point P_0 to the path γ_1 with initial point P_1.

A smooth (=unramified) covering surface of R is a Riemann surface $\tilde{R}$ together with a locally $1-1$ analytic mapping τ of $\tilde{R}$ into R. The mapping τ is called the projection, $P = \tau(\tilde{P})$ the trace of $\tilde{P}$ and $\tilde{P}$ a lift of P. Although τ is part of the notion of covering surface, we will use the notation $\tilde{R}$ rather than $(\tilde{R}, \tau)$ if the projection map τ is fixed.

The covering surface $\tilde{R}$ is called relatively unbounded (=unlimited, regular) if every path γ in R can be lifted to $\tilde{R}$ with an arbitrary initial point $\tilde{P}$ above the initial point P of γ.

Due to the monodromy theorem, a smooth and relatively unbounded covering surface $\tilde{R}$ of R has the property that the projection τ induces an isomorphism of its fundamental group $\tilde{F}$ with base point $\tilde{P}_0$ onto a subgroup G of the fundamental group F with base point $P_0 = \tau(\tilde{P}_0)$.

Conversely, to every such subgroup G there exists an unlimited covering surface $\tilde{R}$ and a point $\tilde{P}_0$ above P_0 such that the fundamental group $\tilde{F}$ with base point $\tilde{P}_0$ projects onto G. The covering surface $\tilde{R}$ is uniquely determined up to a deck homeomorphism (=homeomorphism which leaves the traces invariant).

Let now $\tilde{P}_1 \neq \tilde{P}_0$ be a point of $\tilde{R}$ with the same projection P_0. Then the fundamental group $\tilde{F}_1$ of $\tilde{R}$ with base point $\tilde{P}_1$ projects onto a conjugate subgroup $G_1 = a^{-1} G a$ of G. Here a is the homotopy class of the projection α of a path $\tilde{\alpha}$ joining $\tilde{P}_0$ to $\tilde{P}_1$.

A cover transformation (deck transformation) T of $\tilde{R}$ is a homeomorphism of $\tilde{R}$ onto itself which leaves the traces (projections) invariant. Let $T(\tilde{P}_0) = \tilde{P}_1$. Like every homeomorphism, T induces in a natural way an isomorphism between the two fundamental groups $\tilde{F}$ and $\tilde{F}_1$. As the traces are left invariant, the projections G and G_1 of $\tilde{F}$ and $\tilde{F}_1$ respectively are the same. On the other hand $G_1=a^{-1}Ga$, and therefore a is an element of the normalizer N of G (the subgroup of all elements which commute with G). Conversely, if $a=[\alpha]$ is in the normalizer of G, the lift $\tilde{\alpha}$ of α with initial point $\tilde{P}_0$ has an endpoint $\tilde{P}_1$ above P_0 with a fundamental group which projects onto G. One easily shows that there exists a cover transformation T with $T(\tilde{P}_0)=\tilde{P}_1$. In fact, the group of cover transformations is isomorphic to the factor group N/G. A canonical isomorphism is achieved in the indicated way. If, in particular, G is the identity subgroup of F_{P_0}, the normalizer N of G is equal to F_{P_0} and $\tilde{R}$ is the universal covering surface of R which we denote by $\hat{R}$. Therefore the group of cover transformations of $\hat{R}$ is isomorphic to the fundamental group F of R.

2.2 Annular covering surfaces. Besides the universal covering surface $\hat{R}$, the fundamental group of which projects onto the identity subgroup of F, the following annular covering surfaces will be used (Marden, Richards and Rodin [1], [2]). Let γ be a homotopically non trivial closed path on R with initial point P_0 (i.e. γ is not homotopic to P_0). Its homotopy class $c=[\gamma]$ generates an infinite cyclic subgroup C of F. Let $\tilde{R}$ and $\tilde{P}_0\in\tilde{R}$ above P_0 be such that the fundamental group $\tilde{F}$ with base point $\tilde{P}_0$ projects onto C. As the fundamental group of $\tilde{R}$ is infinite cyclic, it is homeomorphic to a doubly connected plane domain: $\tilde{R}$ is called the annular covering surface induced by γ. We will, in general, identify $\tilde{R}$ with a conformally equivalent annulus $0\leq r_0<|z|<r_1\leq\infty$ of the complex z-plane, together with a projection map τ. As R and $\tilde{R}$ have the same universal covering surface $\hat{R}$, $\tilde{R}$ is the punctured plane $0<|z|<\infty$ if and only if R is parabolic, in the sense that its universal covering surface is $\mathbb{C}$. In the hyperbolic case we can always assume that the outer radius is one, i.e. $\tilde{R}$ is the surface $0\leq r<|z|<1$.

The surface $\tilde{R}$ only depends on the free homotopy class of γ. It is in particular independent of the initial point P_0 on γ. For, let the closed path δ with initial point Q_0 be freely homotopic to γ with initial point P_0. Let h be a deformation which takes P_0 into Q_0 along a path α. The lift $\tilde{\alpha}$ of α with initial. $\tilde{P}_0$ leads to some point $\tilde{Q}_0$, and the continuation $\tilde{\delta}$ of δ from $\tilde{Q}_0$ is closed, as we can lift up h. The path $\tilde{\alpha}$ induces a canonical isomorphism between the fundamental groups $\tilde{F}_{\tilde{P}_0}$ and $\tilde{F}_{\tilde{Q}_0}$. As $\tilde{\delta}$ corresponds to $\tilde{\gamma}$, it generates $\tilde{F}_{\tilde{Q}_0}$. Therefore $\tilde{F}_{\tilde{Q}_0}$ projects onto the subgroup of F_{Q_0} generated by δ, which means that $\tilde{R}$ is the annular covering surface induced by δ, and thus does not depend on the curve in the free homotopy class.

There is another way to find the surface $\tilde{R}$: it is by identification of certain points of the universal covering surface $\hat{R}$ of R. For an arbitrary subgroup $G\subset F_{P_0}$ let $\tilde{R}$ be the corresponding covering surface. The universal covering surface $\hat{R}$ of R also covers $\tilde{R}$ and the group of cover transformations with respect to $\tilde{R}$ is isomorphic to G. If we identify the points of $\hat{R}$ equivalent under this group, we get $\tilde{R}$ back.

In our present case this means the following: choose a point $\hat{P}_0\in\hat{R}$ above P_0. Let $\hat{\gamma}$ be the lift of γ to $\hat{R}$ with initial point $\hat{P}_0$. Let T denote the deck transformation of $\hat{R}$ which takes $\hat{P}_0$ into the endpoint $\hat{P}_1$ of $\hat{\gamma}$. It generates a cyclic subgroup of the group of cover transformations of $\hat{R}$. The surface $\tilde{R}$ is the surface $\hat{R}$ with the points identified which are equivalent under this subgroup. The transformation T corresponds to a parabolic resp. hyperbolic linear transformation of the disk. After a conformal mapping of the disk onto a half plane which takes the fixed points into infinity resp. zero and infinity we can perform the identification in this half plane. The conformal mapping onto a punctured disk in the first case and onto an annulus in the second case is immediate.

As an application we show: let γ be a homotopically non trivial closed curve on an arbitrary Riemann surface R. If $\gamma \sim \gamma^n$, then $n=1$.

Proof. Let $\tilde{R}$ be the annular covering surface associated with γ. We represent $\tilde{R}$ as an annulus $0 \le r_0 < |z| < r_1 \le \infty$ in the z-plane. Let $\tilde{\gamma}$ be the closed lift of γ with initial point z_0 above the arbitrarily chosen initial point P_0 of γ. By definition, $[\tilde{\gamma}]$ generates the fundamental group $\tilde{F}_{z_0}$ of $\tilde{R}$. Therefore its winding number around $z=0$ must be one (with proper orientation of γ). Consequently, the lift of γ^n, which is $\tilde{\gamma}^n$, has winding number n. As the free homotopy $\gamma \sim \gamma^n$ can be lifted to $\tilde{R}$, we get $\tilde{\gamma} \sim \tilde{\gamma}^n$. But this is only possible if the winding numbers of the two curves are the same. Thus $n=1$.

There is a relation between cover transformations of $\tilde{R}$ and closed lifts of γ. Let T be a cover transformation of $\tilde{R}$ with $T(z_0) = z_1 \neq z_0$. Then $T(\tilde{\gamma})$ is a closed lift of γ with initial point z_1. Conversely, assume that γ has a closed lift $\tilde{\gamma}_1$ with initial point $z_1 \neq z_0$ above P_0. Let the winding number of $\tilde{\gamma}_1$ about $z=0$ be n. Then $\tilde{\gamma}_1 \sim \tilde{\gamma}^n$, and by projection $\gamma \sim \gamma^n$ on R. We conclude $n=1$.

Now join z_0 to z_1 by an arc $\tilde{\alpha}$ on $\tilde{R}$. Then $\tilde{\alpha}^{-1}\tilde{\gamma}\tilde{\alpha} \sim \tilde{\gamma}_1$, hence $a^{-1}ca = c$, where a is the homotopy class of the (closed) projection α of $\tilde{\alpha}$ and $c = [\gamma]$. Thus a is commutative with c and hence in the commutator subgroup of C. There exists a cover transformation T with $T(z_0) = z_1$.

2.3 The annular covering surface induced by a Jordan curve. In the sequel we assume γ to be a Jordan curve, $P_0 \in \gamma$. Then the same is true for its lift $\tilde{\gamma}$ with initial point z_0 above P_0. Moreover, $\tilde{\gamma}$ must separate the two boundary circles of $\tilde{R}$, as it generates the fundamental group of $\tilde{R}$.

If R is parabolic, there are only two possibilities: either it is planar, in which case it must be the doubly punctured Riemann sphere with γ separating the two boundary points, or it is the torus.

Let R be hyperbolic. Then the group of cover transformations of $\tilde{R}$ must be a finite rotation group, as a conformal selfmapping of an annulus which interchanges the two boundary components has two fixed points and hence cannot be a cover transformation. We want to show that this group reduces to the identity.

Let ϑ be the angle of rotation of a cover transformation T, and let $z_1 = z_0 \cdot e^{i\vartheta}$, with z_0 the distinguished point above P_0. The Jordan curve $\tilde{\gamma}$ goes over into a Jordan curve $\tilde{\gamma}_1 = T(\tilde{\gamma})$ through z_1 which does not interesect $\tilde{\gamma}$ unless $\vartheta = 0$. Let $\rho_0 = \inf_{z \in \tilde{\gamma}} |z|$, $\rho_1 = \sup_{z \in \tilde{\gamma}} |z|$. Then $\tilde{\gamma}$ has points in common with both circles $|z| = \rho_0$ and $|z| = \rho_1 \ge \rho_0$ and so does $\tilde{\gamma}_1$. We conclude that the two curves intersect, hence $\vartheta = 0$.

It now follows immediately from the reasoning at the end of the preceding section that γ has only one closed lift, which we denote by $\tilde{\gamma}$.

In the same way one sees that γ^n cannot have a closed lift from $z_1 \neq z_0$, for arbitrary n. For otherwise there would be a smallest n with this property. The lift $\tilde{\gamma}^n$ would be a Jordan curve and hence be freely homotopic to $\tilde{\gamma}$. By projection $\gamma \sim \gamma^n$, hence $n=1$, and we are back at the former case.

2.4 Jordan curves homotopic to a point. We are now ready to prove two important topological theorems about Jordan curves on a Riemann surface (Levine [1]; Marden, Richards and Rodin [1]). The proof of the first one only makes use of the universal covering surface, whereas the tool for the second one is the annular covering surface.

Theorem 2.4. *Let γ be a Jordan curve on an arbitrary Riemann surface R which is homotopic to 1. Then it is the boundary of a disk D on R. This disk is uniquely determined except for the case of the Riemann sphere.*

Proof. We assume R to be different from the Riemann sphere. Let $\hat{\gamma}$ be a lift of γ to the universal covering surface $\hat{R}$ of R which we represent in the z-plane. Every such lift is a Jordan curve (it is closed by the monodromy theorem and the lifting itself is bijective). As $\hat{R}$ is a planar surface, $\hat{\gamma}$ is the boundary of a disk $\hat{D}$. We want to show that the restriction of the projection map τ to the closure of $\hat{D}$ is a homeomorphism. Since $\tau(\hat{\gamma})=\gamma$, $D=\tau(\hat{D})$ is then the desired disk bounded by γ.

We first remark that two different lifts $\hat{\gamma}_1$ and $\hat{\gamma}_2$ of a Jordan curve γ cannot intersect. For, let $z \in \hat{\gamma}_1 \cap \hat{\gamma}_2$. Then the projection P of z is a point on γ and as γ is a Jordan curve it corresponds to a uniquely determined value of the parameter of γ. By the uniqueness of the lifting we conclude $\hat{\gamma}_1 = \hat{\gamma}_2$.

Let now z_1 and $z_2 \neq z_1$ be two points in the closure of $\hat{D}$ having the same trace P. There is a cover transformation T of $\hat{R}$ with $T(z_1)=z_2$. In the hyperbolic case ($\hat{R}$ conformally equivalent to the disk) T is either a parabolic or a hyperbolic linear transformation. We represent $\hat{R}$ as a half plane, the fixed points corresponding to infinity or zero and infinity respectively. Let $\hat{D}_1 = T(\hat{D})$. Since one at least of the points z_1 and z_2 must lie in $\hat{D}$, we can assume, by a small simultaneous displacement, that this is the case for both points. Then $\hat{D}$ and $\hat{D}_1$ have the point z_2 in common, and as their boundaries are nonintersecting Jordan curves, one must lie inside the other. This evidently is in contradiction with the geometric nature of T (translation or homothety). In the parabolic case $\hat{R}$ is represented as the complex plane, T is a translation and the reasoning is the same.

2.5 Disjoint, freely homotopic Jordan curves. The next theorem deals with a similar property of a non intersecting pair of freely homotopic Jordan curves.

Theorem 2.5. *Let γ and δ be two non intersecting freely homotopic Jordan curves on an arbitrary Riemann surface R which are not homotopic to a point of R. Then they bound a ring domain D on R. It is uniquely determined except when R is a torus.*

Proof. We first consider the hyperbolic case. Let $\tilde{R}$ be the annular covering surface of R determined by γ (and by δ). It will be represented as an annulus with center zero in the z-plane. Let $\tilde{\gamma}$ and $\tilde{\delta}$ be the closed lifts of γ resp. δ to the surface $\tilde{R}$. They are nonintersecting Jordan curves containing $z=0$ in their interior. Thus they bound a ring domain $\tilde{D}$ situated in $\tilde{R}$. We show, by contradiction, that the projection τ is $1-1$ in the closure of $\tilde{D}$. Let z_1 and z_2 be two points with the same trace. We can assume, without loss of generality, that one of the points, z_1 say, is on the boundary of $\tilde{D}$, while the other is in $\tilde{D}$. For, let both points be in $\tilde{D}$. Join z_1 to $\tilde{\gamma}$ by an arc $\tilde{\alpha}_1$ in $\tilde{D}$. The trace α of $\tilde{\alpha}_1$ can be lifted to z_2 to get an arc $\tilde{\alpha}_2$ the endpoint of which is either in $\tilde{D}$ or outside of $\tilde{D}$. In the second case we shorten α such that its lift $\tilde{\alpha}_2$ is in $\tilde{D}$ except for its endpoint. The endpoint of the continuation of α from z_1 will then be in $\tilde{D}$ (Fig. 1).

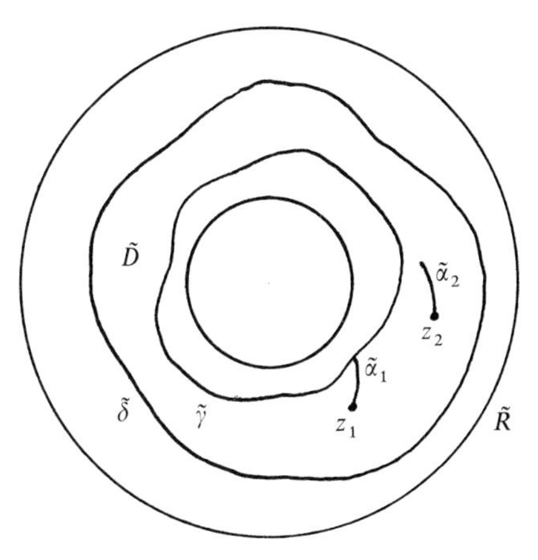

Fig. 1

Now let $z_1 \in \tilde{\gamma}$, $z_2 \in \tilde{D}$. Continue γ from z_2. Since, in the compact set $\bar{\tilde{D}}$ there can only be finitely many points with a given trace, the continuation must close after finitely many iterations. This is a contradiction since (by section 2.3) there cannot be such a closed lift except $\tilde{\gamma}$. Therefore the projection τ is a $1-1$ conformal mapping of $\tilde{D}$ onto a ring domain D which is bounded by γ and δ.

In the parabolic case the underlying surface R is either the doubly punctured Riemann sphere or the torus. In the first case it is a question of plane geometry, as before. So, let R be a torus. Let $\tilde{\gamma}$ and $\tilde{\delta}$ be closed lifts of γ and δ respectively to the annular covering surface $\tilde{R}$ of R defined by γ. If the ring domain $\tilde{D}$ bounded by $\tilde{\gamma}$ and $\tilde{\delta}$ contains two points z_1 and z_2 with the same trace on R, we can again assume that one of the two points is on the boundary of $\tilde{D}$, say $z_1 \in \tilde{\gamma}$. We continue γ from z_2. The lift must close after $n \geq 1$ iterations. Let $\tilde{\alpha}$ be an arc joining z_1 to z_2. Let α be its projection. Then the curve $\alpha\gamma^n\alpha^{-1}$ has a closed lift with initial point z_1 which is either homotopic to z_1 or to $\tilde{\gamma}$. We get, by projection, $ac^n a^{-1}=1$ or $ac^n a^{-1}=c$, hence $c^n=1$ or $c^n=c$, as a commutes with c (the fundamental group is Abelian). We conclude that n must be one and we have found another closed lift $\tilde{\gamma}_1$ of γ which separates $\tilde{\gamma}$ and $\tilde{\delta}$.

We can now repeat the argument and we arrive, after finitely many steps, at a ring domain with the property that the restriction of the projection map to this ring domain is $1-1$.

2.6 Admissible systems of Jordan curves. Certain systems of disjoint Jordan curves on a Riemann surface will play an important role in our later constructions. We call them admissible, for lack of a better name. Moreover, a closed curve which is homotopic to a point is called homotopically trivial.

Definition 2.6. A set of homotopically non trivial Jordan curves γ_i on a Riemann surface R which are mutually disjoint and belong to different free homotopy classes is called an admissible system of Jordan curves.

It is of some interest to determine the maximal number of elements of an admissible system of Jordan curves on a compact surface R or on a compact surface with finitely many punctures. Let R be such a surface, of genus $g \geq 0$, with $r \geq 0$ punctures.

Theorem 2.6. *The maximal number of elements in an admissible system of Jordan curves on a compact Riemann surface R of genus $g \geq 0$ with $r \geq 0$ punctures is $N = 3(g-1) + 2r$, except for the following cases: $g = 0$, $r \leq 1$, where $N = 0$, and $g = 1$, $r = 0$, where $N = 1$.*

Proof. The exceptional cases are easily checked. If $g = 0$ and $r \leq 1$ there evidently is no homotopically non trivial Jordan curve on R. Let $g = 1$, $r = 0$. There are Jordan curves which do not partition the surface. If we cut R along such a curve, we get a cylinder, and there is no other curve possible.

In the general case we proceed by induction with respect to the genus g. (To prove it for $g = 0$ and arbitrary r we apply induction to r.) We also first compute the maximal number of curves in a system where no curve is freely homotopic to an arbitrarily small circle around one of the punctures. We claim that the maximal number is $3(g-1) + r$ in this case (unless $g = 0$, $r \leq 2$, where it is zero, $g = 1$, $r = 0$, where it is one).

Let $g = 0$, $r \geq 3$. Assume that the theorem is true for a Riemann sphere with r punctures. Let $\gamma_1, \ldots, \gamma_p$ be an admissible system of Jordan curves on a Riemann sphere with $r+1$ punctures, satisfying the additional condition mentioned above. Cut the surface along one of the curves, γ_1 say. Let the number of punctures in the two complements of γ_1 be r' and r'' respectively. Evidently $r' \geq 2$, $r'' \geq 2$. If we let the boundary curve γ_1 shrink to a point, the remaining curves in each of the two components satisfy the conditions imposed on the system, with $r'+1$ respectively $r''+1$ punctures, where $r'+1 \leq (r+1) - 2 + 1 = r$ and likewise $r''+1 \leq r$. We therefore can apply the formula to each component, which gives $r'-2$ resp. $r''-2$ as maximal numbers. Thus

$$p \leq (r'-2) + (r''-2) + 1 = r' + r'' - 3 = r - 3 = 3(g-1) + r.$$

It is, on the other hand, easy to give a curve system with $r-3$ elements. Therefore the assertion is proved for $g = 0$.

Let it be proved for some $g \geq 0$; we have to show that it is then also true for $g+1$. Let R be a compact Riemann surface of genus $g+1$ with $r \geq 0$

punctures. Let $\gamma_1, \ldots, \gamma_p$ be a maximal system of Jordan curves satisfying our conditions. At least one of the curves must be non dividing, as we could otherwise add such a non dividing Jordan curve to our system and increase the number of its elements. We cut R along such a curve. The new surface has genus $g-1$ and $r+2$ punctures, after having contracted the two boundary curves which result from the cutting to two points. If $g=1$, $r=0$, this new surface is a sphere with two punctures and thus admits no curves. We get $p=1$, as it should be. In all the other cases we can apply the formula. The remaining system of curves on the new surface satisfies our requirements, hence $p-1 \leq 3(g-2)+r+1$, i.e. $p \leq 3(g-1)+r$. It is, on the other hand, easy to indicate a curve system with $3(g-1)+r$ elements. Thus our assertion is established.

Let now $\gamma_1, \ldots, \gamma_N$ be a maximal admissible system of Jordan curves. For $g=0$, $r \geq 3$ and in all the cases with $g \geq 1$ we must have (up to homotopy) a small circle around every puncture in the system. Hence $N \leq 3(g-1)+2r$. On the other hand, every system of Jordan curves of the type considered can be enlarged by r small circles around the punctures (except for $g=0$, $r=2$), so $N \geq 3(g-1)+2r$, which establishes the theorem.

It is easily seen that every admissible system can be completed to a maximal system.

§ 3. Ring Domains on Riemann Surfaces

3.1 Preliminaries. Let γ be an oriented Jordan curve on a Riemann surface R which is not homotopic to a point. We say that a ring domain $R_0 \subset R$ is associated with γ if a Jordan curve $\gamma_0 \subset R_0$ which separates the two boundary components of R_0 is freely homotopic to γ on R. The modulus of an arbitrary ring domain is defined as $M = \frac{1}{2\pi} \log \frac{r_1}{r_0}$, if $r_0 < |z| < r_1$ is a conformally equivalent annulus. The supremum of the moduli of all the ring domains on R associated with γ is denoted by $M(\gamma)$ and will sometimes be called the modulus of γ.

The doubly connected covering surface of R defined by γ is denoted by $\tilde{R}$. It is represented by an annulus $0 \leq r_0 < |z| < r_1 \leq \infty$ and a projection map τ. The case of the doubly punctured sphere, i.e. $r_0 = 0$ and $r_1 = \infty$, appears if and only if R is parabolic, as R and $\tilde{R}$ have the same universal covering surface. Since γ is a homotopically non trivial Jordan curve on R, this surface is either the doubly punctured sphere or a torus.

Let us now assume that R is hyperbolic. Then $\tilde{R}$ is either an annulus or a punctured disk which can be taken, without loss of generality, to be the unit disk punctured at zero. Because of (2.3), every Jordan curve δ which is freely homotopic to γ has a uniquely determined closed lift $\tilde{\delta}$ on $\tilde{R}$. This extends to ring domains R_0 associated with γ. For, let R_0 be such a domain. For an arbitrary point $P \in R_0$ choose a Jordan curve $\delta \subset R_0$ which contains P and separates the two boundary components of R_0. Let z be the uniquely determined point above P on the closed lift $\tilde{\delta}$ of δ. It evidently does not depend on δ. The mapping $\ell: P \to z = \ell(P)$ is a conformal homeomorphism of R_0 onto a ring domain $\tilde{R}_0$ in $\tilde{R}$; we call it the lifting of R_0. It is the inverse of the restriction of τ to the domain $\tilde{R}_0$.

The modulus of $\tilde{R}$ is denoted by $\tilde{M}(\gamma)$. Because of the existence of liftings we evidently have $M(\gamma) \le \tilde{M}(\gamma)$.

3.2 The reduced modulus. The conformal image of a punctured disk on a Riemann surface is itself called a punctured disk. It has infinite modulus. It is, however, possible to assign it a finite number which is called the reduced modulus (Teichmüller [1]).

Let D be a punctured disk on an arbitrary Riemann surface R, the puncture being a point $P_0 \in R$ (Fig. 2). Let f be a conformal homeomorphism of D

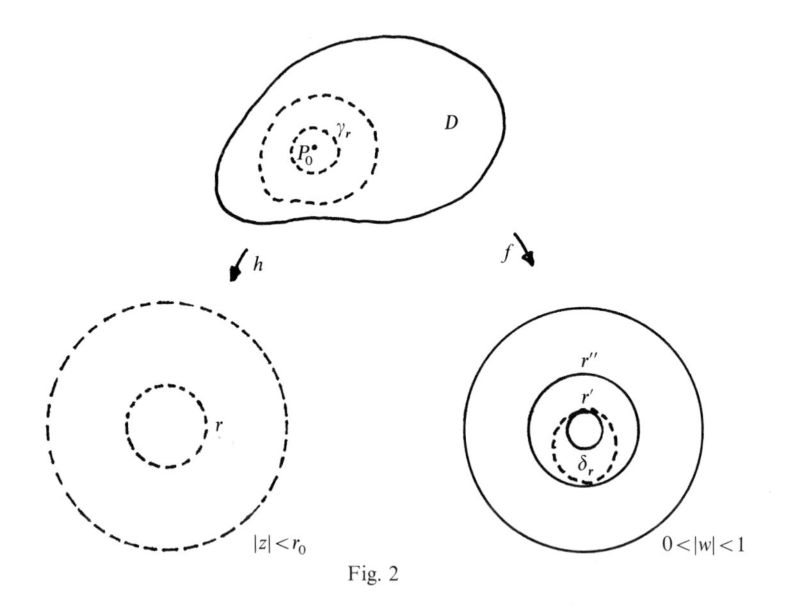

Fig. 2

onto the punctured unit disk $0<|w|<1$, with P_0 corresponding to the point $w=0$. We introduce a conformal parameter $z=h(P)$ near P_0, $|z|<r_0$, $h(P_0)=0$. The circle $|z|=r$ is mapped by h^{-1} onto a curve γ_r which bounds, together with the boundary of D, a ring domain the modulus of which we denote by M_r. The mapping f takes γ_r into a curve δ_r around $w=0$. Let $w=a_1 z + a_2 z^2 + \dots$ be the development of the composition $f \circ h^{-1}$ near $z=0$. Let $r' = \min_{|z|=r} |w(z)|$, $r'' = \max_{|z|=r} |w(z)|$. By monotonicity of the modulus and its conformal invariance we get

$$\frac{1}{2\pi} \log \frac{1}{r''} \le M_r \le \frac{1}{2\pi} \log \frac{1}{r'},$$

hence

$$M_r = \frac{1}{2\pi} \log \frac{1}{r(|a_1| + O(r))}.$$

Therefore the quantity $M_r + \dfrac{1}{2\pi} \log r$ has a limit as $r \to 0$, namely $-\dfrac{1}{2\pi} \log |a_1|$.

Definition 3.2. The reduced modulus of the punctured disk D with respect to the parameter z near the point boundary component P_0, $z=0 \leftrightarrow P_0$, is the limit

$$\dot{M} = \lim_{r \to 0} M_r + \frac{1}{2\pi} \log r,$$

where M_r is the modulus of the ring domain bounded by the curve $|z|=r$ and the boundary component of D which is different from P_0.

We have seen that the reduced modulus is equal to $-\frac{1}{2\pi} \log \left|\frac{dw}{dz}(0)\right|$, where $\frac{dw}{dz}$ is the derivative of a conformal mapping of the punctured disk D onto the punctured unit disk $0<|w|<1$. It depends on the local parameter near P_0. Let $\dot{M} = \dot{M}_z$ be the above reduced modulus with respect to the parameter z. Then, with respect to a parameter $\tilde{z}$ we get

$$\begin{aligned} \dot{M}_{\tilde{z}} &= -\frac{1}{2\pi} \log \left|\frac{dw}{d\tilde{z}}\right| = -\frac{1}{2\pi} \log \left|\frac{dw}{dz}\right| \cdot \left|\frac{dz}{d\tilde{z}}\right| \\ &= \dot{M}_z - \frac{1}{2\pi} \log \left|\frac{dz}{d\tilde{z}}\right|. \end{aligned}$$

Therefore $\dot{M}_z - \frac{1}{2\pi} \log |dz|$ is invariant under a change of parameter.

Let g be a conformal homeomorphism of the punctured disk D onto a disk $0<|\zeta|<\rho$ such that $g'(0)=1$ in terms of the parameter z. Then $f = \frac{g}{\rho}$ is a mapping onto the punctured unit disk. Therefore $\dot{M} = \frac{1}{2\pi} \log \rho$. The number ρ is called the mapping radius of the punctured disk D with respect to the parameter z near P_0. We have seen

Theorem 3.2. *The reduced modulus of a punctured disk D with respect to a parameter z near its point boundary component P_0 is equal to $\frac{1}{2\pi} \log \rho$, where $\rho = \rho_z$ is the mapping radius of D with respect to the parameter z.*

Since $\rho_z = \exp(2\pi \dot{M}_z)$ and $2\pi \dot{M}_z - \log |dz|$ is invariant, $\frac{\rho_z}{|dz|}$ has the same property.

The reduced modulus of a punctured disk D, where the puncture is an ideal pointlike boundary component Γ_0 of R, is defined in the same way. But now the role of the local uniformizer near P_0 is taken over by a conformal mapping h of a neighborhood of Γ_0 onto a punctured disk $0<|z|<r_0$, with Γ_0 corresponding to $z=0$.

The composition $f \circ h^{-1}$ is conformal throughout the disk $|z|<r_0$, which allows us to take over the earlier reasoning.

3.3 Analytic mappings of a punctured disk into a Riemann surface. Let R be an arbitrary open Riemann surface. A nested sequence of domains $G_1 \supset G_2 \supset \dots$ with compact, non empty boundaries, the closures $\bar{G}_i$ of which have empty intersection, defines an ideal boundary component Γ of R. With no loss of generality one can assume that every domain G_i is bounded by a single Jordan curve (Ahlfors and Sario [1], Stoïlow [1]). Two sequences (G_i) and (G'_i) determine the same boundary component, if for every i there is a j such that $G'_j \subset G_i$, and vice versa. The sets G_i are a basis of the set of neighborhoods of Γ. The boundary component Γ', determined by the sequence (G'_i), lies in G_i, if $G'_j \subset G_i$ for all sufficiently large j. A component Γ of the ideal boundary of R is called isolated if there is a G_i of a determining sequence which does not contain any other component Γ'. The boundary component Γ is called planar, if there is a G_i which is conformally equivalent to a plane domain. It is called pointlike, if this plane domain is a punctured disk, the puncture corresponding to Γ.

Theorem 3.3 (Generalized Picard Theorem). *Let f be an analytic mapping of the punctured disk $0<|z|<1$ into a hyperbolic Riemann surface R. Then either zero is a removable singularity of f or else $P=f(z)$ tends to an isolated boundary component Γ of R (Marden, Richards and Rodin [2]).*

In addition one can say that Γ is planar and pointlike (Ohtsuka [1]; Marden, Richards and Rodin [2]).

Proof. For the proof of the first part we follow the presentation of Marden, Richards and Rodin. The supplement is only needed for a special case, namely the projection map of the annular covering surface. It will be proved in Section 3.4.

We make use of the following

Generalized Schottky Lemma. Let f be an analytic mapping of the unit disk $|z|<1$ into a hyperbolic Riemann surface R. Assume that the image $P_0=f(0)$ is contained in a given compact set K which is in turn contained in a fixed open set O. Then there is a positive number r, independent of f, such that $P=f(z)\in O$ for all z, $|z|\leq r$.

The proof is a simple application of Schwarz's lemma. Let $\hat{K}$ be a compact lift of K to the universal covering surface $\hat{R}$ of R. It is easy to get such a set $\hat{K}$, e.g. by covering K by finitely many compact neighborhoods and considering a connected lift of each neighborhood. Moreover, let $\hat{O}$ be an open lift of O, $\hat{O}\supset\hat{K}$. Let $\hat{f}$ be a single valued branch of $\tau^{-1}\circ f$ with $\hat{f}(0)\in\hat{K}$, where τ is the projection map of $\hat{R}$ onto R. Such a branch $\hat{f}$ exists, since the unit disk $|z|<1$ is simply connected. Using the general form of Schwarz's lemma we find $r>0$ such that $\hat{f}(z)\in\hat{O}$ for $|z|\leq r$. Thus $P=f(z)=\tau\circ\hat{f}(z)\in O$.

Picard's theorem is now proved indirectly by lifting f up to the universal covering surface of the punctured disk (which is the half plane $\operatorname{Re}\zeta<0$).

Let there exist a sequence $(z_n)\to 0$ such that $P_n=f(z_n)\to P_0\in R$. Consider an open neighborhood U and a compact neighborhood $K\subset U$ of P_0. Let $r>0$ be the number of the Schottky lemma associated with K and U. The function $g(\zeta)=f(e^\zeta)$ is defined in the half plane $\operatorname{Re}\zeta<0$, hence in the disk $|\zeta-\zeta_n|<-\operatorname{Re}\zeta_n$ with $\zeta_n=\log z_n$ (an arbitrary determination). Its value at ζ_n is $g(\zeta_n)=f(z_n)\in K$ for all sufficiently large n. Therefore $P=g(\zeta)\in U$ as soon as $|\zeta-\zeta_n|\leq r(-\operatorname{Re}\zeta_n)$. For

$-\operatorname{Re}\zeta_n = -\log|z_n| > \frac{\pi}{r}$, i.e. for all sufficiently large n, the vertical stretch $\operatorname{Re}\zeta = \operatorname{Re}\zeta_n$, $|\zeta - \zeta_n| \le \pi$ satisfies the condition. Therefore the circle γ_n: $|z| = |z_n|$ is mapped into U by f.

The image $\gamma_n' = f(\gamma_n)$ is a closed curve in R which is homotopically trivial. Therefore the mapping f has a single valued lift $\hat{f}$ which maps the punctured disk $0 < |z| < 1$ into the universal covering surface $\hat{R}$ of R. Because of the boundedness of $\hat{f}$ the point $z = 0$ is a removable singularity, and as $\hat{f}$ is not a constant, $\hat{f}(0) \in \hat{R}$, $f(0) = \tau \circ \hat{f}(0) = P_0$.

We have shown: If f is not the restriction of an analytic mapping of the disk $|z| < 1$ to the punctured disk, then for every sequence $(z_n) \to 0$ the sequence of points $P_n = f(z_n)$ diverges on R (has no accumulation point on R). Therefore the domains $G_n = f\left(\left\{z; 0 < |z| < \frac{1}{n}\right\}\right)$, $n = 1, 2, \ldots$, define a boundary component Γ of R. It is obviously isolated. It can be shown (Ohtsuka [1]; Marden, Richards and Rodin [2]) that it is also planar and pointlike. In the next section we will give a proof of this supplement in a special case, namely for the projection mapping of the annular covering surface $\tilde{R}$ associated with a Jordan curve γ.

3.4 The case $\tilde{M}(\gamma) = \infty$. Let γ be a homotopically non trivial Jordan curve on a hyperbolic Riemann surface R. Assume that the annular covering surface $\tilde{R}(\gamma)$ has infinite modulus. It can thus be represented by the punctured disk $0 < |z| < 1$ with an analytic projection mapping τ. The curve γ has a unique closed lift $\tilde{\gamma}$ on $\tilde{R}$ which we choose to be positively oriented. Let $\tilde{G}$ be the interior of $\tilde{\gamma}$, punctured at $z = 0$. We claim that the restriction of the projection τ to $\tilde{G}$ is one-one.

Proof. According to section (3.3) the domains

$$G_n = \tau(\tilde{G}_n), \qquad \tilde{G}_n = \left\{z; 0 < |z| < \frac{1}{n}\right\}, \qquad n = 1, 2, 3, \ldots$$

determine an isolated ideal boundary component Γ of R (the curves $\gamma_n = \partial G_n$ are freely homotopie to γ, hence homotopically non trivial; therefore $z = 0$ is not a removable singularity). Let now (D_m) be a nested sequence of domains on R determining the same ideal boundary component Γ and with boundaries $\delta_m = \partial D_m$ which are Jordan curves. We choose m such that $\bar{D}_m \subset G$, then n such that $G_n \subset D_m$. Set $\delta = \delta_m$ and let $P \in \delta$. Since $\delta \subset G \setminus G_n$, there is a point $z \in \tilde{G} \setminus \tilde{G}_n$ above P. The curve δ can be lifted to $\tilde{R}$, starting with z, and the lift closes after finitely many full turns, because it stays in the ring domain bounded by $\tilde{\gamma} = \partial \tilde{G}$ and $\tilde{\gamma}_n = \partial \tilde{G}_n$. Let k be the smallest positive integer such that the lift $\tilde{\delta}^k$ is closed. It must be a Jordan curve which is homotopically non trivial in $\tilde{R}$ and therefore is freely homotopic to $\tilde{\gamma}$. By projection, we conclude $\delta^k \sim \gamma$. It is easy to see that $k = 1$, with proper orientation of δ.

To this end we consider the annular covering surface R' associated with δ. Let δ' be the uniquely determined closed lift of δ on R'. Since δ^k is freely homotopic to γ on R, γ also has a closed lift γ' on R', which is of course a

Jordan curve. Lifting up the homotopy we see that γ' and $(\delta')^k$ are homotopic on R' and therefore have the same winding number about zero. As δ' itself is a Jordan curve on R' we conclude that $k=1$; in particular, $R'=\tilde{R}$. Because of the annulus theorem (2.5), γ and $\delta=\delta_m$ are the two boundary curves of a uniquely determined ring domain R_m on R. Its lift $\tilde{R}_m$ is a ring domain on $\tilde{R}$, bounded by $\tilde{\gamma}$ and $\tilde{\delta}_m$, and the restriction of the projection mapping τ to $\tilde{R}_m$ maps it conformally onto R_m. This is true for every m. As for $m\to\infty$ the curves δ_m recede to Γ, $U\tilde{R}_m$ is the punctured disk $\tilde{G}$. Thus $G=\tau(\tilde{G})$ is a punctured disk on R, bounded by γ and the ideal boundary component Γ, which is therefore pointlike and planar. In particular $M(\gamma)=\infty$.

3.5 Sequences of ring domains

Definition 3.5. A sequence of ring domains R_n on a Riemann surface R is said to converge to a ring domain R_0 if there is a sequence of annuli $0\leq r_n<|z|<s_n\leq\infty$ which converges to an annulus $0\leq r_0<|z|<s_0\leq\infty$ and a sequence of conformal homeomorphisms $g_n\colon r_n<|z|<s_n\to R_n$ which converges locally uniformly in $r_0<|z|<s_0$ to a conformal homeomorphism g_0 of this annulus onto R_0.

Let now γ be an oriented, homotopically non trivial Jordan curve on a hyperbolic Riemann surface R. Let $M(\gamma)<\infty$. The annular covering surface $\tilde{R}$ determined by γ is an annulus $0<r<|z|<1$ together with a projection map τ; we can of course assume that the lift $\tilde{\gamma}$ of γ has winding number one around $z=0$.

Theorem 3.5. *Let the annular covering surface $\tilde{R}(\gamma)$ have finite modulus $\tilde{M}(\gamma)$ (which is equivalent to $M(\gamma)<\infty$). Then every sequence of ring domains R_n associated with γ, the moduli M_n of which do not tend to zero, contains a convergent subsequence.*

Proof. There is a subsequence of the sequence of ring domains, which we denote by (R_n) again, such that the sequence of corresponding moduli M_n converges to a number M_0, $0<M_0\leq M(\gamma)<\infty$. Let g_n be a conformal homeomorphism of the annulus $\rho_n<|\zeta|<1$, $M_n=\dfrac{1}{2\pi}\log\dfrac{1}{\rho_n}$, onto R_n. Denote the mapping which lifts R_n to $\tilde{R}$ by ℓ_n. Then $h_n=\ell_n\circ g_n$ is a conformal homeomorphism of $\rho_n<|\zeta|<1$ onto the lift $\tilde{R}_n$ of R_n, $\tilde{R}_n\subset\tilde{R}$. The sequence (h_n) contains a subsequence (h_{n_i}) which converges locally uniformly in $\rho_0<|\zeta|<1$, $\rho_0=\exp(-2\pi M_0)$, to some analytic function h_0. This function evidently cannot be a constant (the circle $|\zeta|=\sqrt{\rho_0}$, on which the sequence converges uniformly, is mapped onto a Jordan curve of diameter $>2r$ by each h_{n_i}) and therefore is a conformal homeomorphism h_0 of $\rho_0<|\zeta|<1$ onto a ring domain $\tilde{R}_0$. Its composition with the projection τ is a conformal homeomorphism $g_0=\tau\circ h_0$ of $\rho_0<|\zeta|<1$ onto the projection R_0 of $\tilde{R}_0$, and clearly $g_n=\tau\circ h_n\to g_0$ locally uniformly in $\rho_0<|\zeta|<1$.

As an application it follows that there always exists a ringdomain R_0, associated with γ, with largest modulus $M_0=M(\gamma)$.

3.6 Sequences of punctured disks

Theorem 3.6. *Let γ be a Jordan curve on a hyperbolic Riemann surface R which is not homotopic to a point. Let the annular covering surface $\tilde{R}$ of R associated with γ have infinite modulus $\tilde{M}(\gamma)$. Then every sequence of punctured disks R_n associated with γ the reduced moduli M_n of which do not tend to $-\infty$ contains a convergent subsequence.*

Proof. We can assume that the reduced moduli M_n converge to some real number M_0. The lifts $\tilde{R}_n$ of R_n are punctured disks in $0<|z|<1$, the puncture being $z=0$. Let g_n be a conformal mapping of the punctured disk $0<|\zeta|<\rho_n$ onto R_n. We introduce the parameter z near Γ_0 (i.e. the projection map τ is the distinguished local homeomorphism). By proper normalization we can achieve $\frac{d\zeta}{dz}=1$ for every g_n. Then $M_n=\frac{1}{2\pi}\log\rho_n$. Set $M_0=\frac{1}{2\pi}\log\rho_0$. The mappings $h_n=\ell_n\circ g_n$ with ℓ_n the lifting of R_n to $\tilde{R}$ have a subsequence which converges locally uniformly in $|\zeta|<\rho_0$ to a mapping h_0 of $|\zeta|<\rho_0$ onto a simply connected domain $\tilde{D}_0$, $h_0(0)=0$. The projection R_0 of $\tilde{R}_0=\tilde{D}_0\setminus\{0\}$, with $g_0=\tau\circ h_0$ is the desired punctured disk.

Chapter II. Quadratic Differentials

§4. Definition of a Quadratic Differential

Introduction. Every analytic function φ in a domain G of the z-plane defines, in a natural way, a field of line elements dz, namely by the requirement that $\varphi(z)\,dz^2$ is real and positive. This means of course that $\arg dz = -\frac{1}{2}\arg\varphi(z)$ $(\mathrm{mod}\,\pi)$, and thus dz is determined, up to its sign, for every z, where $\varphi(z) \neq 0, \infty$. One may then ask for the integral curves of this field of line elements.

It is desirable to use conformal mapping to study them. Suppose $z = h(\tilde{z})$ is a conformal mapping of a domain $\tilde{G}$ onto G. This transformation takes a line element $d\tilde{z}$ into $dz = h'(\tilde{z})\,d\tilde{z}$. What is the function $\tilde{\varphi}(\tilde{z})$ which determines, in the above sense, the line elements $d\tilde{z}$: "$\tilde{\varphi}(\tilde{z})\,d\tilde{z}^2 > 0$ exactly when $\varphi(z)\,dz^2 > 0$"? If we substitute $z = h(\tilde{z})$ we find

$$\varphi(z)\,dz^2 = \varphi(h(\tilde{z}))\,h'(\tilde{z})^2\,d\tilde{z}^2, \tag{1}$$

and hence

$$\tilde{\varphi}(\tilde{z}) = \varphi(h(\tilde{z}))\,h'^2(\tilde{z}). \tag{2}$$

Any positive multiple of the above $\tilde{\varphi}$ would also define the same field of line elements, but we want to stick to this one. The transformation rule (2), which is the same as that for the square of the derivative of a function, can simply be expressed in the form

$$\tilde{\varphi}(\tilde{z})\,d\tilde{z}^2 = \varphi(z)\,dz^2, \qquad z = h(\tilde{z}). \tag{3}$$

An analytic function φ, together with the transformation rule (3) under a conformal mapping is called a quadratic differential. The maximal integral curves of the field of line elements $\varphi(z)\,dz^2 > 0$ are its trajectories.

If G is a subdomain of the Riemann sphere $\hat{\mathbb{C}}$, and $z = \infty \in G$, one has to apply the inversion $z \to \tilde{z} = \frac{1}{z}$ to get an admissible parameter near $z = \infty$. The transformation rule (2) then gives

$$\tilde{\varphi}(\tilde{z}) = \varphi\left(\frac{1}{\tilde{z}}\right)\frac{1}{\tilde{z}^4}, \qquad \tilde{z} \text{ near zero.} \tag{4}$$

This shows that unlike in the case of a function, where the transformation rule is the invariance (same value at corresponding points), the rule for quadratic differentials leads to a multiplication by $\tilde{z}^{-4}$. The function $\varphi(z) \equiv 1$, considered as a quadratic differential, therefore has a pole of order 4 at $\tilde{z}=\frac{1}{z}=0$. The Riemann sphere is a surface on which one no longer has a single complex parameter which is valid all over the surface. This is even more so on an arbitrary Riemann surface, where there exist only local parameters.

An analytic function φ in a domain $G\subset\mathbb{C}$ also defines, in a natural way, a direction field, namely by the requirement $\varphi(z)\,dz>0$, i.e. $\arg dz=-\arg\varphi$ $(\operatorname{mod} 2\pi)$. The integral curves now have an orientation. (The transformation rule for φ is the same as for the derivative of a function.) The lack of orientation is one of the reasons why quadratic differentials are more important in geometric function theory than linear differentials: they allow for more singularities as well as more complicated behaviour in the large.

Quadratic differentials appear in Schiffer's work in connection with schlicht mappings (Schiffer [1]) and, independently, in Teichmüller's famous papers [1], [2] and [3]. In [1], they are connected with the solution of moduli problems, and in [3] with extremal quasiconformal mappings, where the trajectories are the lines of largest stretching. Teichmüller laid the basis to a general theory. More detailed discussions are found, subsequently, in Schaeffer and Spencer [1] and Jenkins and Spencer [1], who treat rational quadratic differentials in the plane. The first systematic account of quadratic differentials on finite Riemann surfaces is in Jenkins [3].

Consequent use of conformal mapping is made in Strebel [7], Pfluger [1] and Jensen [1]. A survey of the theory is given in Strebel [16]. More references will follow at the relevant places in the course of our treatment.

In this chapter, some basic definitions and facts concerning quadratic differentials and arbitrary Riemann surfaces are given. A more detailed discussion of the trajectory structure and the metric induced by a quadratic differential follows in the subsequent chapters.

4.1 Definition 4.1. Let R be a Riemann surface with a given conformal structure $\{(U_\nu, h_\nu)\}$. A (meromorphic) quadratic differential φ on R is a set of meromorphic function elements φ_ν in the local parameters $z_\nu=h_\nu(P)$ for which the transformation law

$$\varphi_\nu(z_\nu)\,dz_\nu^2=\varphi_\mu(z_\mu)\,dz_\mu^2, \qquad dz_\mu=\frac{dz_\mu}{dz_\nu}\,dz_\nu,$$

holds whenever z_μ and z_ν are parameter values which correspond to the same point P of R. The quadratic differential is called holomorphic, if all the φ_ν are holomorphic.

The function element φ_ν is also called the representation of the quadratic differential φ in terms of the parameter z_ν.

In a similar way one defines differentials of the first order (linear differentials) or of any order n.

Let the pair (U_0, h_0) be compatible with the given conformal structure on R. Then we can define a meromorphic function element φ_0 of the parameter z_0 in the open set $V_0=h_0(U_0)$ by the requirement

$$\varphi_0(z_0)\,dz_0^2=\varphi_\nu(z_\nu)\,dz_\nu^2.$$

By the invariance property this is independent of the particular representation φ_ν which we choose if a point $P \in U_0$ lies in different neighborhoods of the conformal structure. In particular, the function elements of φ can be defined in the maximal structure on R which is equivalent to the given structure $\{(U_\nu, h_\nu)\}$. Two quadratic differentials φ and ψ on a Riemann surface R are called equal if all their corresponding function elements on this maximal structure are equal. As a consequence of the above remark it is sufficient if $\varphi_\nu = \psi_\nu$ for a set (U_ν, h_ν) which covers R (and therefore determines the conformal structure). Actually, by analytic continuation, $\varphi_\nu = \psi_\nu$ for corresponding elements in some (fixed) parameter z_ν is also sufficient.

If we keep a certain parameter fixed, we often call it z instead of z_ν and write $\varphi(z)\,dz^2$ instead of $\varphi_\nu(z_\nu)\,dz_\nu^2$.

It is clear from the definition that the quadratic differentials on a given Riemann surface form a linear space over the complex numbers.

4.2 Regular points, critical points. While it clearly does not make sense to speak of the value of a quadratic differential φ at a point $P \in R$ (since it depends on the local parameter near P), it does make sense to speak of the zeroes and poles of φ. For, suppose z, w are local parameters near P, with $\varphi_1(z)$ and $\varphi_2(w)$ the corresponding representations of φ. We assume, for convenience of notation, that $P \leftrightarrow z = 0 \leftrightarrow w = 0$, and therefore $z = c_1 w + c_2 w^2 + \dots$, $c_1 \neq 0$. Then, if $\varphi_1(z)$ has a zero (pole) of order n resp. $-n$ at $z = 0$, so does $\varphi_2(w)$ at $w = 0$.

To prove it, let $\varphi_1(z)$ have an expansion about $z = 0$ of the form

$$\varphi_1(z) = z^n(a_n + a_{n+1} z + \dots), \qquad a_n \neq 0.$$

Then, by the transformation rule,

$$\begin{aligned}\varphi_2(w) = \varphi_1(z)\left(\frac{dz}{dw}\right)^2 &= \varphi_1(c_1 w + c_2 w^2 + \dots)(c_1 + 2c_2 w + \dots)^2 \\ &= (c_1 w + c_2 w^2 + \dots)^n\,[a_n + a_{n+1}(c_1 w + \dots) + \dots]\,(c_1 + 2c_2 w + \dots)^2 \\ &= c_1^{n+2} a_n w^n + \text{higher powers of } w \\ &= w^n(b_n + b_{n+1} w + \dots), \qquad \text{with } b_n = c_1^{n+2} a_n \neq 0.\end{aligned}$$

Observe that $a_n = b_n$ iff $c_1^{n+2} = 1$, which is true for every change of the local parameter if and only if $n = -2$. On the other hand, the order of a zero or a pole is always invariant.

Definition 4.2. The critical points of a quadratic differential φ are its zeroes and poles. All other points of R are called regular points of φ. A holomorphic point is either a regular point or a zero. Poles of the first order and zeroes will be called finite critical points, poles of order greater or equal to two infinite critical points.

4.3 The lift of a quadratic differential. Let $f: \hat{R} \to R$ be a holomorphic mapping of a Riemann surface $\hat{R}$ into a Riemann surface R. The surface $\hat{R}$ together with the mapping f is called a covering surface (possibly branched) of R. Let $\hat{P}$ and $P=f(\hat{P})$ be corresponding points. Choose local parameters $\hat{z}$ near $\hat{P}$ and z near P and assume that the mapping has the representation $z=f(\hat{z})$ in terms of these parameters. $\hat{P}$ is called a branching point (ramification point) of order n, $n=1,2,\ldots$, of the covering surface $(\hat{R}, f)$ if it is a point of multiplicity of order $n+1$ for f.

Let φ be a quadratic differential on R. We define $\hat{\varphi}(\hat{z})$ by the equation

$$\hat{\varphi}(\hat{z})\, d\hat{z}^2 = \varphi(z)\, dz^2.$$

For a given parameter $\hat{z}$ this evidently does not depend on the parameter z because of the transformation law for φ. On the other hand, if we change $\hat{z}$, $\hat{\varphi}(\hat{z})$ is transformed in the correct manner. Therefore the above equation defines a quadratic differential $\hat{\varphi}$ on $\hat{R}$. It is called the lift of the quadratic differential φ to the covering surface $(\hat{R}, f)$.

4.4 Reflection to the double. Under certain conditions it is possible to continue a quadratic differential on a bordered Riemann surface to its double. Let φ be a quadratic differential on a bordered surface R with border Γ and assume that $\varphi(z)$ is real on some open interval $\Gamma_\lambda \subset \Gamma$ in terms of a specified local parameter z. The local homeomorphisms along the border of R have been chosen to map it onto intervals of the real axis. If we admit more general homeomorphisms which map the corresponding border intervals onto analytic Jordan arcs, we have to postulate that $\varphi(z)\, dz^2$ is real along the border, i.e. for tangential dz, in terms of these parameters. Now let U be an arbitrary open set of the conformal structure of R with U^* its mirror image on R^*. Let z be a local parameter in U; then $z^* = \bar{z}$ is a local parameter in U^*. We define

$$\varphi^*(z^*) = \overline{\varphi(z)} = \overline{\varphi(\overline{z^*})}.$$

This is an analytic function element of its variable z^*. If z corresponds to a point $P \in \Gamma_\lambda$, we have $z = x = z^*$. The function element φ is defined in an upper half neighborhood of x, the function element φ^* in a lower half neighborhood. As $\varphi(x)$ is real, $\varphi^*(x) = \varphi(x)$. Both function elements coincide on the real axis and hence are restrictions of an analytic function element in the symmetric neighborhood: We have a quadratic differential on the whole symmetric surface, i.e. the double $\tilde{R}$ of R with respect to the set of open boundary intervals Γ_λ along which $\varphi(z)\, dz^2$ is real and along which R has been welded to its mirror image R^*.

4.5 Examples. 4.5.1. The quadratic differential $\varphi = 0$ evidently exists on every Riemann surface R.

4.5.2. Let R be a domain G of the complex z-plane $\mathbb{C}$. The conformal structure is defined by one open set, namely G, and the identity mapping. An arbitrary analytic function $\varphi(z)$ in G can be considered as a quadratic differential on R.

4.5.3. Let R be the Riemann sphere. It can be covered by two open sets, $U_1 = \mathbb{C}$ and $U_2 = \mathbb{C} \cup \{\infty\} \setminus \{0\}$. As parameters we introduce z and $w = \frac{1}{z}$. We can give an arbitrary function $\varphi_1(z)$ which is defined in the whole plane and compute the function element $\varphi_2(w)$ by the transformation rule $\varphi_2(w)\,dw^2 = \varphi_1(z)\,dz^2$. We get

$$\varphi_2(w) = \varphi_1\left(\frac{1}{w}\right) \cdot \frac{1}{w^4}.$$

If φ_1 is a rational function, the corresponding quadratic differential φ has at most poles and is therefore also called rational. But as we see from the above transformation law, the degree (= sum of the orders of its zeroes minus the sum of the orders of its poles) is equal to -4. E.g. there is a quadratic differential on R with $\varphi_1(z) \equiv c \neq 0$. It has a pole of order 4 at infinity.

4.5.4. To find the quadratic differentials φ on the torus R we consider its universal covering surface. It is conformally equivalent to the complex plane $\mathbb{C}$. The projection map f of $\mathbb{C}$ onto R is doubly periodic. Let z be a parameter in a neighborhood of a point $P_0 \in R$ and denote by $\varphi(z)$ the corresponding function element. Let w_0 be a point which is projected onto P_0. Then the lift $\hat{\varphi}$ of φ in the neighborhood of w_0 is

$$\hat{\varphi}(w) = \varphi(z)\left(\frac{dz}{dw}\right)^2.$$

If a is an arbitrary period, in the neighborhood of $w_0 + a$ we have

$$\hat{\varphi}(w+a) = \varphi(z)\left(\frac{dz}{d(w+a)}\right)^2 = \hat{\varphi}(w).$$

Thus $\hat{\varphi}$ is a doubly periodic function in the plane. If, in particular, φ is assumed to be holomorphic, $\hat{\varphi}$ must be a constant.

Conversely, let $\hat{\varphi}$ be an arbitrary doubly periodic meromorphic function in the plane (with the proper periods). For a local parameter z, $z = f(w)$, we define

$$\varphi(z) = \hat{\varphi}(w)\left(\frac{dw}{dz}\right)^2.$$

This is uniquely determined because of $\hat{\varphi}(w+a) = \hat{\varphi}(w)$ and $\dfrac{d(w+a)}{dz} = \dfrac{dw}{dz}$. Moreover, for a change of the parameter z it obeys the correct transformation rule. A holomorphic quadratic differential on a torus has no zero (its degree is zero). It is the square of a linear differential. The linear space of all holomorphic quadratic differentials on a torus has complex dimension one.

§ 5. The Function $\Phi(P) = \int \sqrt{\varphi(z)}\,dz$

5.1 Distinguished parameter near regular points. For a differential ψ of the first order ($\psi(z)\,dz$ invariant under any change of the parameter) an important quantity is its integral $\Psi(z) = \int \psi(z)\,dz$. It is a function on the surface R (and could therefore be written in the form $\Psi(P)$, $P \in R$), but of course only determined up to an arbitrary additive constant and in general not single valued on the whole surface.

In order to get an invariant integral in the case of a quadratic differential φ, we first pass to a linear differential, by taking the square root, then integrate:

$$w=\Phi(z)=\int\sqrt{\varphi(z)}\,dz.$$

Every regular point P_0 of φ has a neighborhood in which a single valued branch of this function can be chosen (by integrating one of the two single valued branches of $\sqrt{\varphi(z)}$. The same would be true in a neighborhood of a zero or a pole of even order.) We may speak of the value $\Phi(P)$ of the function Φ at a point P of the surface. But in general we introduce a local parameter z and consider its representation $\Phi(z)$ in terms of this parameter. For any two determinations $\Phi_1(z)$ and $\Phi_2(z)$ near the same regular point we evidently have

$$\Phi_2(z)=\pm\Phi_1(z)+\text{const}.$$

A sufficiently small neighborhood U of a regular point P_0 is mapped homeomorphically by a branch Φ_0 onto an open set V in the w-plane. We can introduce w as a conformal parameter in U_0. The differential dw becomes

$$dw=\Phi'(z)\,dz=\sqrt{\varphi(z)}\,dz,$$

therefore by squaring

$$dw^2=\varphi(z)\,dz^2.$$

This means that the quadratic differential has, in terms of this parameter, the representation $\equiv 1$. If $\tilde{w}$ is another parameter near P_0 with this property, we have, because of $dw^2=d\tilde{w}^2$,

$$\tilde{w}=\pm w+\text{const}.$$

Theorem 5.1. *In a neighborhood of every regular point P_0 of φ we can introduce a local parameter w, in terms of which the representation of φ is identically equal to one. The parameter is given by the integral $w=\Phi(z)=\int\sqrt{\varphi(z)}\,dz$. It is uniquely determined up to a transformation $w\to\pm w+\text{const}$ and it will be called the distinguished or natural parameter near P_0.*

5.2 Maximal φ-disks. Let P_0 be a regular point of φ. Let Φ_0 be a single valued branch of Φ near P_0, $\Phi_0(P_0)=0$. There is a neighborhood U of P_0 which is mapped $1-1$-conformally onto an open set V in the w-plane. We can assume, by restriction, that V is a disk around $w=0$. The inverse Φ_0^{-1} is a conformal homeomorphism of V into the surface R and there evidently is a largest open disk V_0 around $w=0$ in which the analytic continuation of Φ_0^{-1} (which is still denoted by Φ_0^{-1}) is homeomorphic (Fig. 3). The image $U_0=\Phi_0^{-1}(V_0)$ is called the largest φ-disk around P_0; its φ-radius r_0 is the Euclidean radius of V_0. U_0 and r_0 evidently do not depend on the branch of Φ which we have chosen. Let $r_0=\infty$ for some point P_0. Then Φ_0^{-1} maps the complex plane $\mathbb{C}$ onto the maximal φ-disk U_0 around P_0. Because of Section 3.3, $w=\infty$ either corresponds to a point $P_\infty\in R$ or to an isolated, pointlike ideal boundary component. In the first case, $R=U_0\cup\{P_\infty\}$, i.e. R is conformally equivalent to the Riemann sphere, whereas in the second case R evidently is conformally equivalent to the complex plane. It can then of course be completed by adding one point P_∞. We

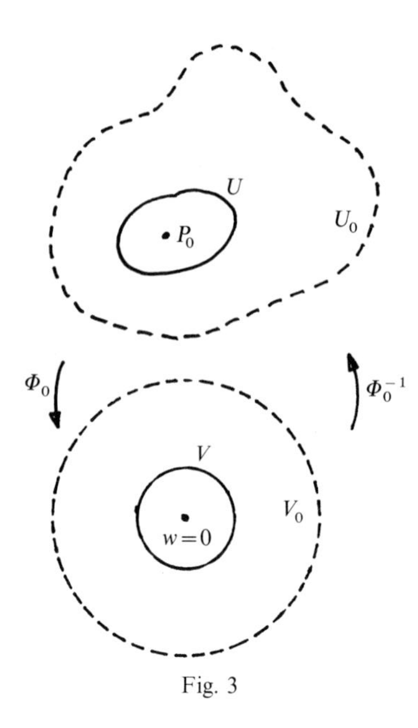

Fig. 3

can introduce the parameter $\tilde{w}=\frac{1}{w}$ near this point. The representations of the quadratic differential in terms of the parameters w and $\tilde{w}$ are $\varphi_1(w)=1$, $\varphi_2(\tilde{w}) =\frac{1}{\tilde{w}^4}$.

Clearly, if $r_0=\infty$ for one point P_0, it is so for every regular point of the surface. If it is finite, it is a continuous function of the point P_0. For, let P be the point on R which corresponds to the value $w_0+\Delta w$, with Φ_0 a fixed branch in U_0, $\Phi_0(P_0)=w_0$, and $|\Delta w|<r_0$. Then the φ-radius $r(P)$ at P satisfies the double inequality

$$r_0-|\Delta w|\leq r(P)\leq r_0+|\Delta w|.$$

For $P\to P_0$, $\Delta w\to 0$ and therefore $r(P)\to r_0$. Summing up we have the following

Definition 5.2. Let φ be a meromorphic quadratic differential on an arbitrary Riemann surface R. A φ-disk is a region which is mapped homeomorphically onto a disk in the complex plane by a branch of Φ.

Theorem 5.2. *Every regular point P is the center of a uniquely determined largest φ-disk. Its φ-radius $r(P)$, i.e. the Euclidean radius of the corresponding disk in the plane, is either identically equal to infinity or else it is a continuous function of P. In the first case, R is the complex plane or the Riemann sphere and the quadratic differential is rational with a single fourth order pole.*

5.3 The metric associated with a quadratic differential. By a path on a Riemann surface R we mean a mapping of an interval of the real axis into R. It makes sense of course to speak of a continuous path, a differentiable path or a locally rectifiable path etc., because these notions do not depend on the local parameter on R. A regular or smooth path is continuously differentiable with a non vanishing derivative.

A curve or arc on R is a class of paths with monotonic parameter transformations of the corresponding class allowed, either monotonically increasing or decreasing. The curve has an orientation, if only monotonically increasing transformations are admitted.

Let γ be a rectifiable curve in a disk U_0 around a regular point P_0. It is mapped by a branch Φ_0 of Φ onto a curve $\gamma'\subset V_0=\Phi_0(U_0)$. The (Euclidean) length of γ' does not depend on the branch of Φ which we have chosen in U_0. An arbitrary rectifiable curve γ on R which does not go through any critical point of φ can be subdivided into intervals each one of which lies in a disk. Then the length of the image of each of these intervals is defined and it is easy to see that the total length does not depend on the subdivision. (Equivalently, we can continue a branch of Φ along γ and thus get an image γ' of the whole curve, the Euclidean length of which is still the same number.) This length can also be computed by means of the differential $dw=\sqrt{\varphi(z)}\,dz$ in terms of an arbitrary local parameter on R. We get

$$|\gamma|_\varphi=\int_{\gamma'}|dw|=\int_\gamma|\varphi(z)|^{1/2}\,|dz|.$$

As the integrand is positive this also makes sense if γ goes through critical points of φ; but it becomes infinite whenever γ contains an infinite critical point (pole of order ≥ 2).

Definition 5.3. The differential $|dw|=|\varphi(z)|^{1/2}\,|dz|$ is called the length element of the φ-metric, the metric associated with the quadratic differential φ. The length of a curve γ in this metric is denoted by $|\gamma|_\varphi$ and called its φ-length.

The corresponding area element is $|\varphi(z)|\,dx\,dy$; it is evidently also invariant under a change of parameter. The total area of the Riemann surface R in this metric is the L^1-norm of φ:

$$|R|_\varphi=\|\varphi\|=\iint_R|\varphi(z)|\,dx\,dy.$$

An isolated singularity of φ has a neighborhood with finite φ-area if and only if it is a first order pole. (Therefore a meromorphic quadratic differential on a compact Riemann surface has finite norm iff it has only finite critical points.) The necessity follows easily from Cauchy's formula:

Let $P\in R$ be an isolated singularity of φ, with z, $z=0\leftrightarrow P$, a local parameter near P. Let φ have the Laurent series

$$\varphi(z)=\sum_{-\infty}^{\infty}a_n z^n$$

in terms of z. Cauchy's integral representation gives, for sufficiently small $r>0$,

$$a_n=\frac{1}{2\pi i}\int_{|z|=r}\frac{\varphi(z)}{z^{n+1}}\,dz,$$

hence

$$2\pi |a_n| r^{n+1} \leq \int_0^{2\pi} |\varphi(r e^{i\vartheta})| r \, d\vartheta.$$

If P has a neighborhood with finite φ-area, integration shows immediately that $a_n = 0$ for $n \leq -2$.

On the other hand, if $\varphi(z) = \sum_{-1}^{\infty} a_n z^n$, $|\varphi(z)| \leq |a_{-1}| \cdot r^{-1} + \text{const.}$, and therefore P has a neighborhood with finite φ-area.

It is also easy to see that a finite critical point can be joined with any neighboring point by a curve of finite φ-length, whereas every such connection has infinite length in case of a pole of order ≥ 2.

5.4 Shortest curves in the neighborhood of a regular point. The φ-length of a curve is most easily measured in the plane of the distinguished parameter $w = \Phi(z)$, where it is the Euclidean length. Let U_0 be the largest φ-disk around a regular point P_0, with r_0 its radius. Choose a branch Φ_0 of Φ in U_0, $\Phi_0(P_0) = 0$. Let V be the disk $|w| < r = \frac{r_0}{2}$, $U = \Phi_0^{-1}(V)$. Let $P_i \in U$, $w_i = \Phi_0(P_i)$, $i = 1, 2$. Then the image of the straight line segment joining w_1 and w_2 in the w-plane is the unique shortest curve (in the φ-metric) joining P_1 and P_2 on the surface (Fig. 4). This is evident, because it is true in U_0, and any curve joining P_1 and P_2 and leaving U_0 has length $> 2r$. We have seen:

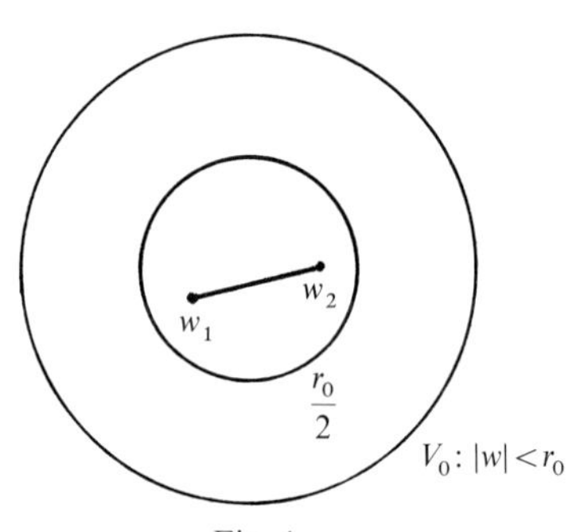

Fig. 4

Theorem 5.4. *Every regular point P_0 of φ has a neighborhood U such that any two points P_1 and P_2 in U can be joined by a uniquely determined shortest arc (in the φ-metric). It stays in U and its tangential elements dz satisfy*

$$\arg \varphi(z) \, dz^2 = \arg dw^2 = \text{const.} \pmod{2\pi}.$$

5.5 Geodesics. Trajectories

Definition 5.5.1. Let $\gamma\colon t \to P = \gamma(t)$, $a < t < b$, be a locally rectifiable curve. It is called a geodesic if it is locally shortest, which means that every point t is contained in the interior of an interval $[t_1, t_2]$ such that the arc $\gamma([t_1, t_2])$ is the shortest connection of the two points $P_1 = \gamma(t_1)$ and $P_2 = \gamma(t_2)$ with respect to the underlying metric.

Every locally rectifiable curve γ which does not pass through a pole of φ is of course locally rectifiable in the φ-metric and can be parametrized by the arc length in this metric. The same is evidently true for a geodesic. A parameter which measures the arc length in the φ-metric (i.e. for $u_1<u_2$ the difference u_2-u_1 is equal to the length of the subarc corresponding to the interval $[u_1,u_2]$) is called a natural parameter of γ. If u and u' are two natural parameters, there is of course a transformation $u'=\pm u+\text{const}$ such that $\gamma(u')\equiv\gamma(u)$. In the following we will always use this parametrization.

Definition 5.5.2. A straight arc (with respect to the quadratic differential φ) is a smooth curve γ along which

$$\arg dw^2=\arg\varphi(z)\,dz^2=\theta=\text{const.},\qquad 0\le\theta<2\pi.$$

This implies in particular that $\varphi(z)\neq 0,\ \infty$ on γ. In other words a straight arc only contains regular points of φ.

A maximal straight arc is a straight arc which is not properly contained in another one. We say that of two locally rectifiable curves γ_1 and γ_2 one is properly contained in the other, say $\gamma_1\subset\gamma_2$, if, in the natural parameter with equal normalization, the mapping γ_1 is a restriction of the mapping γ_2.

We speak, in particular, of a horizontal arc, if $\theta=0$, and of a vertical arc, if $\theta=\pi$. For an arbitrary θ we sometimes speak of a θ-arc.

Definition 5.5.3. A horizontal trajectory or simply a trajectory of the quadratic differential φ, is a maximal horizontal arc. A maximal vertical arc is called a vertical trajectory. A maximal straight arc $\arg\varphi(z)\,dz^2=\theta$ is sometimes called a θ-trajectory.

The local existence and uniqueness of a horizontal arc through every regular point P follows from conformal mapping, by means of Φ. The global existence and uniqueness of a trajectory is then proved in the usual way by continuation (see also Chapter IV).

Theorem 5.5. *Let φ be a meromorphic quadratic differential on an arbitrary Riemann surface R. Then through every regular point of φ there exists a uniquely determined trajectory.*

In particular, two trajectories never have a common point, unless they coincide. On the other hand, a trajectory can be periodic. We can then restrict the parameter to a primitive period and we speak of a closed trajectory.

It is clear that the straight θ-arcs of φ are the horizontal arcs of the quadratic differential $e^{-i\theta}\varphi$. In particular, the vertical trajectories of φ are the trajectories of $-\varphi$. It is therefore sufficient to consider the horizontal trajectories of quadratic differentials. We will however see that the geodesics are in general composed of straight arcs of different inclinations, the vertices being finite critical points.

Remark. The inequality $\varphi(z)\,dz^2>0$ defines an invariant line element at every regular point of φ: Let $\arg\varphi(z)=\vartheta\ (\operatorname{mod}2\pi)$, then $\arg dz=-\vartheta/2\ (\operatorname{mod}\pi)$. The maximal integrals of these line elements are the trajectories of φ. More accurately: The maximal solutions of the differential equation $\varphi(z)\left(\frac{dz}{du}\right)^2=1$ are the trajectories of φ in the natural parameter u.

Because of the square dz^2, the trajectories of a quadratic differential do not have a natural orientation, unlike the trajectories of a linear differential. This is a reason for their much wider geometric applicability.

Definition 5.5.4. Let α: $u \to P=\alpha(u)$, $-\infty \leq u_1 < u < u_2 \leq \infty$ be a non closed trajectory in its natural parametrization. Then, for $u_0 \in (u_1, u_2)$, the restriction of α to one of the subintervals $(u_1, u_0]$, $[u_0, u_2)$ is called a trajectory ray with initial point $\alpha(u_0)=P_0$. Rays will usually be denoted by the symbols α^- and α^+ respectively.

Chapter III. Local Behaviour of the Trajectories and the φ-Metric

Introduction. This chapter is devoted to the investigation of the trajectory structure and the φ-metric near the critical points. The treatment in Schaeffer and Spencer [1] and Jenkins [3] is based on the theory of differential equations. Here, special conformal parameters will be introduced in terms of which the representation of the quadratic differential becomes particularly simple. This is achieved by computing the integral Φ and expressing it in simple terms. Differentiation and squaring then gives the representation of φ with respect to the new parameters. The procedure is particularly simple if there are no logarithmic terms, which is always the case if n is positive, or negative and odd (Strebel [7], Pfluger [1], Jensen [1]).

§ 6. Distinguished Parameter Near Critical Points

6.1 Critical points of odd order. Let P_0 be a critical point of the quadratic differential φ of odd order n and let its representation in terms of a given parameter z near P_0, $P_0 \leftrightarrow z=0$, be

$$\varphi(z)=z^n(a_n+a_{n+1}z+\dots), \qquad a_n \neq 0.$$

In a sufficiently small neighborhood of zero we can select a single valued branch of the square root of the term inside the parenthesis, say

$$(a_n+a_{n+1}z+a_{n+2}z^2+\dots)^{1/2}=b_0+b_1z+b_2z^2+\dots.$$

We then get for $\sqrt{\varphi}$ (locally outside $z=0$)

$$\sqrt{\varphi(z)}=z^{n/2}(b_0+b_1z+b_2z^2+\dots),$$

and by integrating term by term

$$w=\Phi(z)=z^{\frac{n+2}{2}}(c_0+c_1z+c_2z^2+\dots),$$

with

$$c_k=\frac{2b_k}{n+2(k+1)}.$$

Let

$$d_0+d_1z+d_2z^2+\dots=(c_0+c_1z+c_2z^2+\dots)^{\frac{2}{n+2}}$$

be a single valued branch of the right hand side in some sufficiently small neighborhood of $z=0$. Then

$$w=\Phi(z)=\zeta^{\frac{n+2}{2}},$$

with

$$\zeta=z(d_0+d_1 z+\dots),$$

which is a conformal homeomorphism of some neighborhood of $z=0$ onto a disk $|\zeta|<\rho$.

The representation of the quadratic differential φ in terms of the parameter ζ is found by differentiating Φ and squaring. We get

$$dw=\Phi'(\zeta)\,d\zeta=\frac{n+2}{2}\,\zeta^{n/2}\,d\zeta$$

and therefore

$$dw^2=\left(\frac{n+2}{2}\right)^2\zeta^n\,d\zeta^2$$

which is again single valued in the full neighborhood. Let $\tilde{\zeta}$, $\tilde{\zeta}=0\leftrightarrow\zeta=0$, be another parameter with the above representation. We have

$$\frac{n+2}{2}\,\zeta^{n/2}\,d\zeta=\frac{n+2}{2}\,\tilde{\zeta}^{n/2}\,d\tilde{\zeta},$$

and integration leads to

$$\tilde{\zeta}^{\frac{n+2}{2}}=\zeta^{\frac{n+2}{2}}+C,$$

hence

$$\tilde{\zeta}=[\zeta^{\frac{n+2}{2}}+C]^{\frac{2}{n+2}}.$$

For $n\geq -1$ we have $\tilde{\zeta}(0)=C^{\frac{2}{n+2}}=0$ and thus $C=0$. But for $n\leq -3$, n odd this must also be true, as otherwise

$$\tilde{\zeta}=\zeta[1+C\cdot\zeta^{-\frac{n+2}{2}}]^{\frac{2}{n+2}}$$

is not a single valued function near $\zeta=0$. Therefore

$$\tilde{\zeta}=c\,\zeta,\qquad \text{with } c=\exp\left(\frac{k}{n+2}\cdot 2\pi i\right),\qquad k=0,1,\dots,n+1.$$

It is obvious that one can also get a representation of φ of the form $A\zeta^n d\zeta^2$, with any $A\neq 0$, in particular one of the form $\zeta^n d\zeta^2$, but the above representation is simpler for the integral Φ.

Theorem 6.1. *Let P_0 be a critical point of the quadratic differential φ with odd exponent n. Then there is a local parameter ζ near P_0, $P_0\leftrightarrow\zeta=0$, such that the representation of φ in terms of this parameter is*

$$\varphi(\zeta)\,d\zeta^2=\left(\frac{n+2}{2}\right)^2\zeta^n\,d\zeta^2.$$

(The integral $\Phi(\zeta)$ then has the simple form $\Phi(\zeta)=\zeta^{\frac{n+2}{2}}+\text{const.}$) The parameter ζ is uniquely determined up to a factor $c=\exp\left(\frac{k}{n+2}2\pi i\right)$, $k=0,1,\dots,n+1$.

6.2 Zeroes of even order. At a zero of even order (and of course also at a regular point) P_0 we have, putting $n=2m$,

$$\sqrt{\varphi(z)}=z^m(b_0+b_1 z+\dots), \qquad m=0,1,2,\dots$$

and therefore

$$w=\Phi(z)=z^{m+1}(c_0+c_1 z+\dots), \qquad c_k=\frac{b_k}{m+k+1}.$$

Setting

$$\zeta=z(c_0+c_1 z+\dots)^{\frac{1}{m+1}}=z(d_0+d_1 z+\dots)$$

we get

$$w=\zeta^{m+1}+\text{const.}$$

and

$$dw^2=(m+1)^2\,\zeta^{2m}\,d\zeta^2=\left(\frac{m+2}{2}\right)^2\zeta^n\,d\zeta^2.$$

The representation of φ is the same; the uniqueness of the parameter up to a transformation $\zeta\to c\cdot\zeta$, $c=\exp\left(\frac{k}{n+2}2\pi i\right)$, $k=0,1,\dots,n+1$, is proved as before, using $\zeta=0\leftrightarrow\tilde{\zeta}=0$. Summing up 6.2 and the relevant part of 6.1 we have

Theorem 6.2. *In the neighborhood of any finite critical point P_0 we can introduce a local parameter ζ, $P_0\leftrightarrow\zeta=0$, in terms of which the quadratic differential has the representation*

$$\varphi(\zeta)\,d\zeta^2=\left(\frac{n+2}{2}\right)^2\zeta^n\,d\zeta^2.$$

The parameter is uniquely determined up to a factor $c=\exp\left(\frac{k}{n+2}\cdot 2\pi i\right)$, $k=0,1,\dots,n+1$. It is called the distinguished or natural parameter near P_0.

6.3 Poles of order 2. In this case $(m=-1)$, the leading term after the integration is logarithmic, namely

$$w=\Phi(z)=b_0\log z+b_1 z+\frac{b_2}{2}z^2+\dots.$$

This can be made $=b_0\log\zeta$ by setting

$$\log\zeta=\log z+\frac{b_1}{b_0}z+\frac{b_2}{2b_0}z^2+\dots,$$

i.e.

$$\zeta=z\cdot\exp\left(\frac{b_1}{b_0}z+\frac{b_2}{2b_0}z^2+\dots\right)=z(1+d_1 z+\dots).$$

The representation of the quadratic differential in terms of ζ becomes

$$dw^2=\left(\frac{b_0}{\zeta}\,d\zeta\right)^2=\frac{a_{-2}}{\zeta^2}\,d\zeta^2.$$

The coefficient a_{-2} is the invariant leading coefficient of φ at P_0. The uniqueness of the parameter ζ up to an arbitrary constant factor $c \neq 0$, $\tilde{\zeta} = c \cdot \zeta$, follows readily.

Theorem 6.3. *At a pole P_0 of order 2 the quadratic differential φ can be represented in the form*

$$dw^2 = \left(\frac{a_{-2}}{z^2} + \frac{a_{-1}}{z} + a_0 + a_1 z + \dots\right) dz^2 = \frac{a_{-2}}{\zeta^2} d\zeta^2.$$

The parameter ζ, $P_0 \leftrightarrow \zeta = 0$, is uniquely determined up to an arbitrary factor $c \neq 0$.

6.4 Poles of even order ≥ 4. Let $m \leq -2$, i.e. $n = 2m = -4, -6, \dots$. Integration of the (single valued) square root

$$\sqrt{\varphi(z)} = z^m (b_0 + b_1 z + \dots)$$

now also produces a logarithmic term of the form $b \log z$, $b = b_{|m|-1}$. We get

$$w = \Phi(z) = z^{m+1}(c_0 + c_1 z + \dots) + b \log z.$$

Let $\zeta = z(d_0 + d_1 z + \dots) = z\, e^{h(z)}$, $d_0 \neq 0$ be such that

$$w = \zeta^{m+1} + b \log \zeta + c. \tag{1}$$

We then get the representation (for the square root of φ)

$$\begin{aligned} dw = \sqrt{\varphi(z)}\, dz &= \left((m+1)\, \zeta^m + \frac{b}{\zeta}\right) d\zeta \\ &= \left(\frac{n+2}{2} \zeta^{n/2} + \frac{b}{\zeta}\right) d\zeta. \end{aligned} \tag{2}$$

From (1) we deduce

$$g(z) = e^{(m+1)h(z)} + z^{-(m+1)}(b\, h(z) + c), \tag{3}$$

with $g(z) = c_0 + c_1 z + \dots$, and by differentiation

$$\begin{aligned} g'(z) + (m+1)\, z^{-(m+2)}(b\, h(z) + c) \\ = ((m+1)\, e^{(m+1)h(z)} + z^{-(m+1)} b)\, h'(z) \end{aligned} \tag{4}$$

with $m = -2, -3, \dots$. This is a differential equation for the function h which is solvable in a neighborhood of $z = 0$ with any initial value $h(0)$. Let h be a solution. By integration we get (3), up to an additive constant, which must be zero if $h(0)$ satisfies

$$c_0 = e^{(m+1)h(0)}.$$

Therefore (2) holds. The distinguished parameter is not uniquely determined because (2) holds with any value of c in (1). Instead of a differential equation one can apply the implicite function theorem to solve (3) (see Jensen [1]). By the transformation $\zeta = A\tilde{\zeta}$ we can get an arbitrary coefficient of the first term in (2), but not of the second, which is always b.

Theorem 6.4. *Let P be a pole of φ of even order ≥ 4. Then it is possible to introduce a parameter ζ in some neighborhood of P, $P \leftrightarrow \zeta = 0$, in terms of which a branch of the square root of φ has the representation*

$$\sqrt{\varphi(z)}\, dz = \left(\frac{n+2}{2} \zeta^{n/2} + \frac{b}{\zeta}\right) d\zeta.$$

§ 7. Trajectory Structure Near Critical Points

In order to find the local trajectory structure, we express φ in terms of the distinguished parameter. The structure in an arbitrary parameter plane then is a conformal image of the given picture.

7.1 Finite critical points. Let P_0 be a finite critical point, i.e. a zero or a first order pole. In fact, the regular points do not have to be excluded here, which means that we just assume $n \geq -1$. The representation of the quadratic differential in terms of the distinguished parameter ζ is

$$\left(\frac{n+2}{2}\right)^2 \zeta^n d\zeta^2$$

and the integral of the square root of φ becomes

$$w = \Phi(\zeta) = \zeta^{\frac{n+2}{2}},$$

choosing the constant of the integration to be zero. The full angle $0 \leq \arg \zeta \leq 2\pi$ is subdivided into $n+2$ equal sectors,

$$\frac{2\pi}{n+2} k \leq \arg \zeta \leq \frac{2\pi}{n+2}(k+1), \qquad k = 0, 1, \ldots, n+1.$$

Every sector is mapped by Φ onto an upper or lower halfplane. The pre-images of the horizontals $\operatorname{Im} w = \text{const}$ are the horizontal arcs with respect to φ. We have the following pictures (Figs. 5, 6, 7):

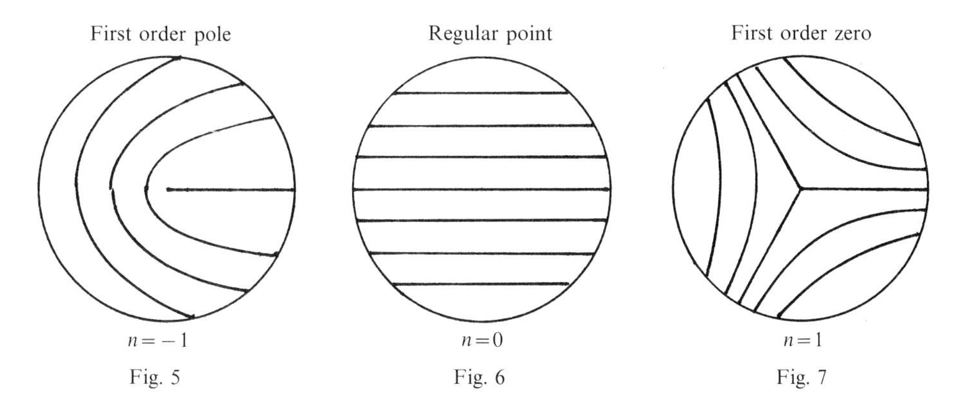

Fig. 5 Fig. 6 Fig. 7

Theorem 7.1. *Let P be a finite critical point and ζ the distinguished parameter near P. Subdivide the disk $|\zeta|<\rho$ for some suitable $\rho>0$ by the radii $\arg\zeta = k\frac{2\pi}{n+2}$, $k=0,\ldots,n+1$ into $n+2$ sectors. Map each one of the sectors onto a half circle in the upper or lower half plane by means of the function $w=\zeta^{\frac{n+2}{2}}$. The trajectory arcs are the lines which are mapped into the horizontals. In particular, the distinguished radii are the critical trajectory arcs ending at (or emerging from) P.*

In case of a first order pole we have of course just one sector.

7.2 Poles of order two. In terms of the distinguished parameter ζ the quadratic differential has the representation

$$\frac{a_{-2}}{\zeta^2}\,d\zeta^2,$$

which gives $w=\Phi(\zeta)=b\log\zeta$, $b=\sqrt{a_{-2}}$. We have three cases to distinguish.

$a_{-2}>0$ (b real, Fig. 8): The universal covering surface of the punctured disk $0<|\zeta|<\rho$ is mapped onto the half plane $\operatorname{Re} w<b\log\rho$ if $b>0$ (resp. $\operatorname{Re} w>b\log\rho$ if $b<0$). The horizontals in the w-plane correspond to the radii in the ζ-plane which are therefore the trajectory arcs of φ near $\zeta=0$.

$a_{-2}<0$ (b imaginary, Fig. 9): The universal covering surface of $0<|\zeta|<\rho$ is mapped onto an upper or a lower half plane. The horizontals in the w-plane correspond to the circles around $\zeta=0$. The trajectories in the neighborhood of $\zeta=0$ are closed curves surrounding $\zeta=0$.

a_{-2} not real (Fig. 10): The universal covering surface of $0<|\zeta|<\rho$ is mapped onto a halfplane with a boundary which is neither vertical nor horizontal. The trajectory arcs are logarithmic spirals tending to (or emerging from) $\zeta=0$.

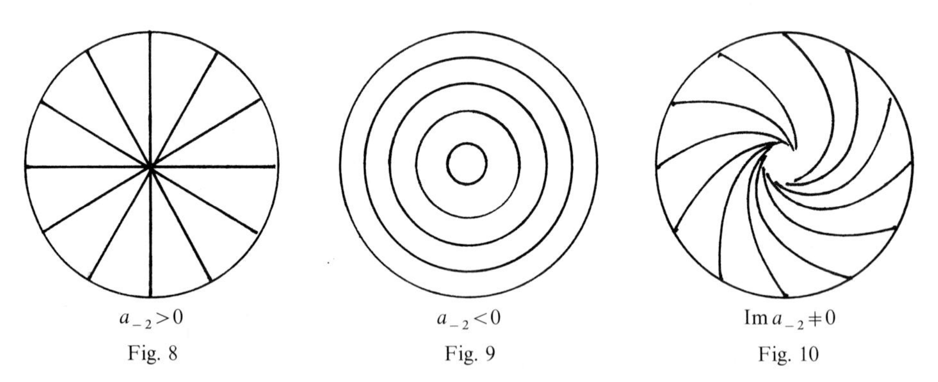

$a_{-2}>0$ Fig. 8 | $a_{-2}<0$ Fig. 9 | $\operatorname{Im} a_{-2}\neq 0$ Fig. 10

Theorem 7.2. *In the neighborhood of a pole P of order two the trajectory arcs in terms of a distinguished parameter ζ are the images of the horizontal straight lines (half lines) by the mapping $w\mapsto\zeta=\exp\frac{w}{\sqrt{a_{-2}}}$. They are the circles around $\zeta=0$ if $a_{-2}<0$, the radii if $a_{-2}>0$ and logarithmic spirals in all the other cases.*

The trajectory structure in the plane of an arbitrary parameter near P results from the above structure, as always, by a conformal mapping.

7.3 Poles of order ≥3 without logarithmic term. Let P be a pole of odd order ≥3 or of even order ≥4 with vanishing logarithmic term. Then, according to Theorems 6.1 and 6.4, the quadratic differential has, in terms of the distinguished parameter ζ, the representation

$$\varphi(z)\,dz^2 = \left(\frac{n+2}{2}\right)^2 \zeta^n d\zeta^2.$$

We get $w=\zeta^{\frac{n+2}{2}}$, up to an additive constant, which we choose to be zero. Let $\tilde{\zeta} = \frac{1}{w}$. Then the horizontals in the w-plane correspond to the circles through $\tilde{\zeta}=0$ with the real axis as tangent. By the mapping $\zeta=\tilde{\zeta}^{\frac{2}{|n|-2}}$ we find the trajectories in terms of the distinguished parameter ζ. We get the following pictures for $|\zeta|<\rho$ (Figs. 11, 12, 13):

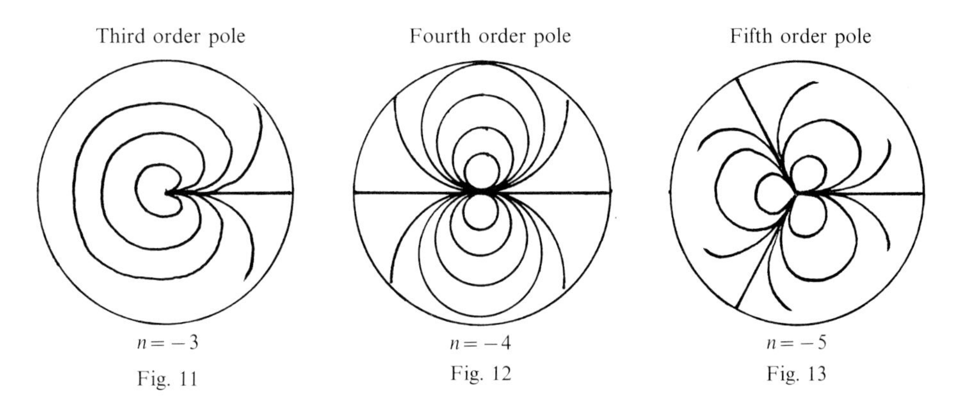

Fig. 11 Fig. 12 Fig. 13

In particular, all the trajectories tend to P in one of $|n|-2$ directions which bound congruent sectors with angles $\frac{2\pi}{|n|-2}$. It has to be noticed, however, that we do not have any invariantly distinguished trajectories here. We read off that P has a neighborhood with the property that every trajectory ray which enters it tends to P.

7.4 Poles of even order ≥4. The representation of φ in terms of a distinguished parameter is, according to Section 6.4,

$$dw=\sqrt{\varphi(z)}\,dz=\left((m+1)\,\zeta^m+\frac{b}{\zeta}\right)d\zeta$$

with $m=\frac{n}{2}=-2,-3,\ldots$. In order to recognize the trajectory structure we introduce the new variable $z=\zeta^{m+1}$. We get

$$dw = \left(1+\frac{b'}{z}\right) dz, \qquad b' = \frac{b}{m+1}.$$

Choose $r>0$ such that in $U=\{z=x+iy, |x|+|y|>r\}$ we have $\left|\frac{b'}{z}\right|<\frac{1}{2}$. Then, by Theorem 5.5, through every point of U there is a unique oriented trajectory α in the natural parametrization, i.e. a solution $z(u)$ of the differential equation

$$1 = \left(1+\frac{b'}{z}\right)\frac{dz}{du}.$$

Unless α ends on ∂U in one of its directions, it tends to $+\infty$ for $u\to\infty$ and to $-\infty$ for $u\to-\infty$. This is the case for every trajectory which cuts the imaginary axis outside of the interval $|y|\leq r$. Assume now that α tends to infinity for $u\to+\infty$. Then, as $1+\frac{b'}{z}\to 1$ for $z\to\infty$, the argument of its tangent vector tends to zero. Therefore, for any fixed angle $\vartheta>0$, it will stay in the sector $|\arg z|<\vartheta$ for all sufficiently large u. An analogous statement holds for $u\to-\infty$. We now map the neighborhood U of $z=\infty$ onto a neighborhood $\tilde{U}$ of $\tilde{z}=0$ by the inversion $\tilde{z}=\frac{1}{z}$. In $\tilde{U}$, every trajectory which does not meet $\partial\tilde{U}$ tends to zero in the sector $|\arg\tilde{z}|<\vartheta$ for $u\to\infty$, for any $\vartheta>0$, and likewise, for $u\to-\infty$, in the sector $|\arg\tilde{z}-\pi|<\vartheta$. Moreover, the argument of the tangent, taken as an angle in the interval $\left[-\frac{\pi}{2},\frac{\pi}{2}\right]$, tends to zero. The corresponding statement is true for a trajectory ray (half a trajectory) which starts at a point of $\partial\tilde{U}$. Finally we pass to the plane of the distinguished parameter ζ by $\zeta=\tilde{z}^{\frac{-1}{m+1}} =\tilde{z}^{\frac{2}{|n|-2}}$. The picture is approximately the same as in the case without a logarithmic term. We have proved

Theorem 7.4. *Let P be a pole of order ≥ 3, i.e. $n\leq -3$. Then there are $|n|-2$ directions at P forming equal angles, and the trajectories enter in these distinguished directions into P. There is a neighborhood U of P such that every trajectory ray which enters into U tends to P. The two rays of any trajectory which stays in U tend to P in two consecutive distinguished directions.*

§ 8. The φ-Metric Near Finite Critical Points

8.1 Shortest curves near a zero of φ. Let P be a zero of order n of the quadratic differential φ. The representation, in terms of the distinguished parameter ζ, is

$$\varphi(\zeta)\,d\zeta^2 = \left(\frac{n+2}{2}\right)^2 \zeta^n d\zeta^2, \qquad |\zeta|<\rho$$

and therefore Φ becomes, as a function of ζ,

$$w=\Phi(\zeta)=\zeta^{\frac{n+2}{2}}.$$

Theorem 8.1. *Let P be a zero of φ of order n. Then there is a neighborhood $U=U(P)$ such that any two points P_1 and P_2 in U can be joined by a unique shortest arc γ. It stays in U and it is either a straight arc* $\arg dw^2=\arg\varphi(z)\,dz^2$ $=\text{const.}$ *or else it is composed, in terms of ζ, of two radii, enclosing angles* $\geq\frac{2\pi}{n+2}$ (Figs. 14, 15). *We call this the angle condition for the geodesics at the zeroes of φ. (It should be noticed that, in the second case,* $\arg dw^2$ *can be different along the two radii.)*

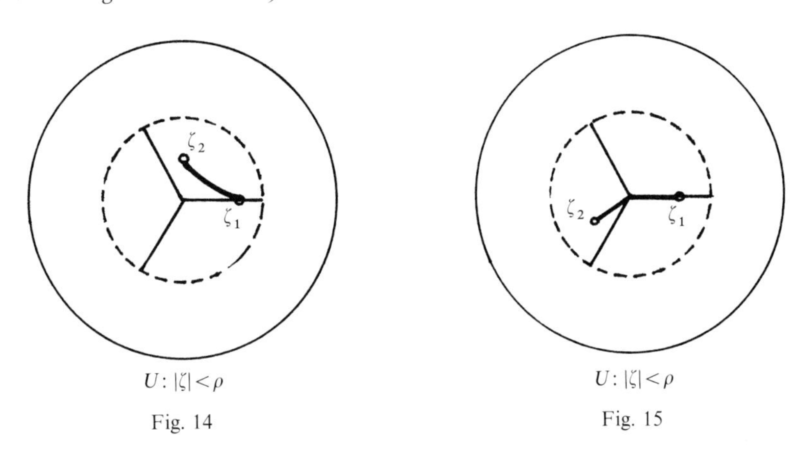

U: $|\zeta|<\rho$

Fig. 14

U: $|\zeta|<\rho$

Fig. 15

Proof. The φ-length of the radius of the circle $|\zeta|=\rho$ is equal to $\rho^{\frac{n+2}{2}}$. We choose $\rho_1=\rho(\frac{1}{2})^{\frac{2}{n+2}}$. Then any two points ζ_1, ζ_2 in U: $|\zeta|<\rho_1$ can be joined by an arc (namely the two radii) which is shorter than any arc which leaves the disk $|\zeta|<\rho$.

Let γ be an arbitrary arc joining ζ_1 and ζ_2 in U. We may assume, without loss of generality, that $\zeta_1\neq 0$. First, let $|\arg\zeta-\arg\zeta_1|\leq\frac{2\pi}{n+2}$ for any $\zeta\in\gamma$, $\zeta\neq 0$. This means that the arc γ lies in the two (closed) sectors which are mapped onto closed half planes by Φ and which are adjacent to the radius through ζ_1. The curve $\gamma'=\Phi(\gamma)$ joins the two points $w_1=\Phi(\zeta_1)$ and $w_2=\Phi(\zeta_2)$ in the w-plane and its Euclidean length is clearly greater or equal to that of the rectilinear interval joining the two points. The statement of the theorem can be read off immediately for this case.

Let now $\zeta_2=0$, and let $\zeta\in\gamma$ be the first point where γ leaves one of the adjacent sectors. Then the subarc $[\zeta_1,\zeta]$ of γ is longer than the sum of the radii with endpoints ζ_1 and ζ respectively and therefore the entire curve γ is longer than the radius to ζ_1, unless it is equal to this radius. This argument also immediately shows that any γ which passes through zero is longer than the sum of the two radii unless it is equal to this sum. If ζ_2 is in the interior of the union of the two adjacent sectors, the straight arc joining ζ_1 and ζ_2 is still shorter; in the other case, every curve γ contains a point ζ where it first leaves the two adjacent sectors and is thus minorized by the union of the two radii, which proves the theorem.

8.2 Shortest curves in the neighborhood of a first order pole. Let P be a pole of the first order of φ. Then, in terms of the distinguished parameter ζ, φ has the representation

$$dw^2 = \varphi(\zeta)\, d\zeta^2 = \tfrac{1}{4}\zeta^{-1}\, d\zeta^2, \qquad |\zeta| < \rho,$$

and the integral Φ is equal to

$$w = \Phi(\zeta) = \zeta^{\frac{1}{2}}.$$

Any two points ζ_1 and ζ_2 in U: $|\zeta| < \rho_1 = \rho/4$ can be joined by a curve, namely the two radii, which is shorter than any connection which leaves $|\zeta| < \rho$. In order to find the shortest curve we may assume, without loss of generality, that $\zeta_1 \neq 0$ and $\arg \zeta_1 = 0$. We look at the two sheeted branched covering surface of $\sqrt{\zeta}$ and cut the upper sheet, say, along the positive real axis, whereas ζ_1 is supposed to lie in the lower sheet. It is mapped, by Φ ($\arg \Phi(\zeta_1) = 0$), onto the full disk slit along the negative real axis. The φ-length of any curve is equal to the Euclidean length of its Φ-image.

Let us first consider a curve γ which joins ζ_1 with ζ_2 without passing through zero. We fix $\arg \zeta$ for $\zeta \in \gamma$, $\zeta \neq 0$, by continuity, starting with $\arg \zeta_1 = 0$, and set $\arg \Phi(\zeta) = \frac{1}{2} \arg \zeta$. Let $\zeta_0 \in \gamma$ be the first point such that $|\arg \zeta| = 2\pi$ (if there is such a point). Let $r_i = |\zeta_i|^{\frac{1}{2}}$, $i = 0, 1$ (i.e. the φ-length of the radius with endpoint ζ_i). Then the φ-length of the subarc γ_1 of γ with endpoints ζ_1 and ζ_0 is $\geq r_0 + r_1$, with equality only if it is equal to the sum of the two radii. The same argument works, if ζ_0 is the first point on γ which is equal to zero. Therefore the radius $[0, \zeta_1]$ is the unique shortest connection between ζ_1 and zero.

The same argument shows that among all curves γ which join ζ_1 with ζ_2, $\zeta_2 \in U$ arbitrary, and which pass through zero, the sum of the two radii is the unique shortest. And it also shows that the unique shortest connection of two points ζ_1 and ζ_2 which are on the same radius is the radial interval $[\zeta_1, \zeta_2]$.

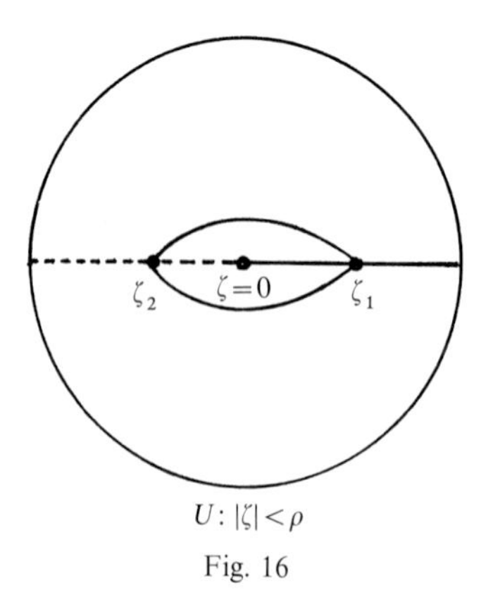

U: $|\zeta| < \rho$

Fig. 16

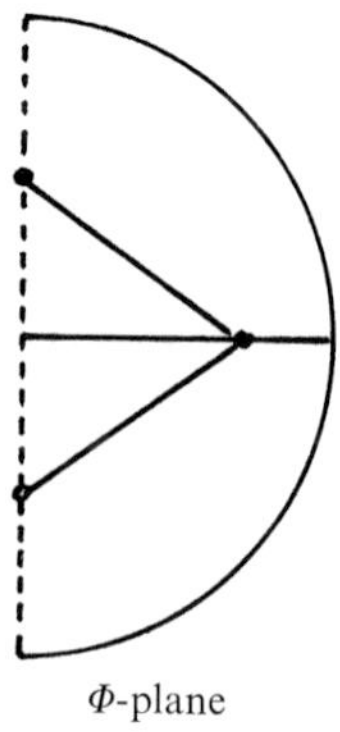
Φ-plane

Fig. 17

Let now ζ_2 be on the negative real axis. Then there are two straight arcs $\arg dw = \text{const.}$ joining ζ_1 with ζ_2, as is readily seen in the Φ-plane. It is also easy to recognise that they are shortest arcs, of equal length (Figs. 16, 17). This lack of uniqueness is the reason why we puncture the disk at $\zeta = 0$ and only compare curves in the same homotopy class of the punctured disk. This class is determined by the winding number $\Delta \arg \zeta = \arg \zeta_2 - \arg \zeta_1$.

Theorem 8.2. *Let $|\Delta \arg \zeta| < 2\pi$. Then there is a unique shortest curve γ_0 in the homotopy class, namely the straight arc connecting the two points. If $|\Delta \arg \zeta| \geq 2\pi$, every connecting arc of the class is longer than the sum of the two radii, which can then be considered as the unique shortest limit of the curves of the class.*

Proof. To prove the first part of the theorem, let us first assume that $|\arg \zeta - \arg \zeta_1| < 2\pi$ for all points $\zeta \in \gamma$. Then γ is mapped continuously by Φ onto a curve γ' which joins $w_1 = \Phi(\zeta_1)$ with $w_2 = \Phi(\zeta_2)$ which is evidently longer than the straight Euclidean arc γ'_0 joining the two points, unless $\gamma' = \gamma'_0$.

If $|\arg \zeta - \arg \zeta_1| \geq 2\pi$ for some ζ, let ζ_0 be the first such point. Then the subarc γ_1 with endpoints ζ_1 and ζ_0 is longer than the sum of the two radii $[0, \zeta_1] + [0, \zeta_2]$. If we replace γ_1 by this sum, we get a new curve $\tilde{\gamma}$ which is shorter than γ and which meets zero. It is thus longer than the sum of the two radii $[0, \zeta_1]$ and $[0, \zeta_2]$. Moreover, this sum is, by Euclidean geometry, strictly longer than the straight arc between the two points.

To prove the second part we must simply realize that, in this case, there always is a point $\zeta \in \gamma$ with $|\arg \zeta - \arg \zeta_1| = 2\pi$. The earlier reasoning shows that the sum of the radii $[0, \zeta_1]$ and $[0, \zeta_2]$ is a minorant for all curves of the gives class.

Chapter IV. Trajectory Structure in the Large

§9. Closed Trajectories

9.1 The representation of the trajectories by Φ^{-1}. In Section 5.5 the trajectories of φ were defined to be the maximal horizontal arcs. Locally they are the images, by a branch of the analytic mapping Φ^{-1}, of the horizontal intervals in the w-plane. We are now going to represent the trajectory α through a regular point P_0 in the large by this mapping. In this manner, we get α in its natural parametrization; moreover, as Φ^{-1} will be defined in a neighborhood of α, it also describes the relation between α and the neighboring trajectories.

Let φ be a meromorphic quadratic differential on an arbitrary Riemann surface R. Consider a regular point P_0 of φ and let U_0 be the maximal φ-disk with center P_0. We fix a branch Φ_0 of Φ in U_0 with $\Phi_0(P_0)=0$. Pick a point $u_1 \in V_0 = \Phi_0(U_0)$ on the real axis (Fig. 18). The point $P_1 = \Phi_0^{-1}(u_1)$ is a regular

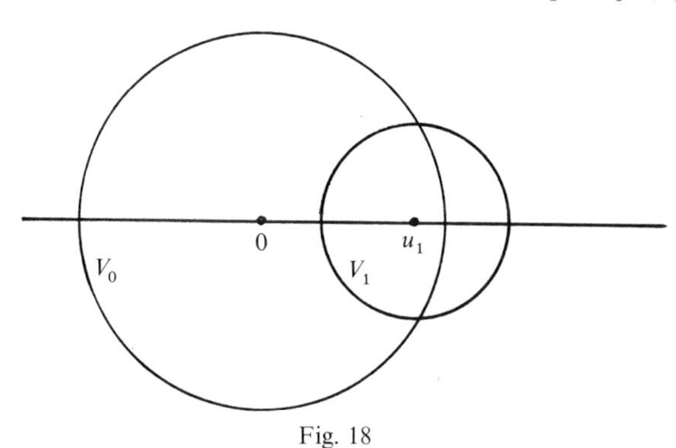

Fig. 18

point of φ. Let U_1 be the maximal φ-disk around P_1. We choose the branch Φ_1 of Φ in U_1 which is equal to Φ_0 in the intersection $U_0 \cap U_1$. Then Φ_1^{-1} is the direct analytic continuation of Φ_0^{-1}. Picking a point u_2 on the real axis in the disk $V_1 = \Phi_1(U_1)$ we continue in the same way. We get a finite chain $C = V_0 \cup V_1 \cup \ldots \cup V_k$ of disks in the w-plane with centers on the real axis and a mapping of C into the surface R, which is locally $1-1$ and conformal. Let D be the union of all these chains. If $w \in D$, $w \in C$ for some chain C. We define $\Phi^{-1}(w)$ as the value at w of the analytic continuation of Φ_0^{-1} along this chain. As the intersection of two chains is connected and contains V_0, Φ^{-1} is uniquely defined in D.

Let $\Delta = D \cap \mathbb{R}$. Then $\Phi^{-1}(u)$, $u \in \Delta$ defines the trajectory α through P_0 in the natural parametrization. For, if $u \in \Delta$, $u \in V$ for some disk $V = \Phi(U)$, image by a branch of $\int \sqrt{\varphi(z)}\,dz$ of a φ-disk. Therefore $0 < du^2 = \varphi(z(u))\,dz^2$. Let I be a closed subinterval of α which contains P_0. Then it can be covered by finitely many φ-disks with centers on I, U_0 being among them. Choosing the proper branch Φ_0 in U_0, I evidently corresponds to a subinterval of Δ. Thus α is maximal.

9.2 Closed trajectories. Let there be two points u_0 and u_1 on Δ which are mapped by Φ^{-1} onto the same point $P_0 \in R$. Then, as Φ^{-1} is locally homeomorphic, there are two points with minimal distance in the interval $[u_0, u_1]$ which have the same property. We call them again u_0 and $u_1 > u_0$. Let V_0 and V_1 be the maximal open disks with centers u_0 and u_1 resp. in which Φ^{-1} is univalent. They are pre-images of the maximal φ-disk U_0 with center P_0. The two respective branches $\Phi_0\colon U_0 \to V_0$ and $\Phi_1\colon U_0 \to V_1$ of Φ satisfy the relation $\Phi_1 = \pm\Phi_0 + \text{const.}$ The minus sign cannot hold, because otherwise there would be two points u_0' and u_1' in (u_0, u_1) which are images of the same point P_0' by Φ_0 and Φ_1 respectively. Therefore $\Phi_1 = \Phi_0 + a$, with $a = u_1 - u_0$. From this we conclude that $\Phi^{-1}(w+a) = \Phi^{-1}(w)$ for all $w \in V_0$, and the identity continues to hold under analytic continuation. We get $\Delta = \mathbb{R}$, the domain D has the period a (allows horizontal shifts by multiples of a) and Φ^{-1} is a periodic function in D with a as primitive period. The interval $[u_0, u_1)$ is mapped by Φ^{-1} onto the whole image set, and as $\Phi^{-1}(u_0) = \Phi^{-1}(u_1)$, this image is a Jordan curve. We therefore also speak of closed trajectories instead of periodic trajectories.

9.3 Embedding of a closed trajectory in a ring domain. Let the closed trajectory α_0 be the $1-1$ image of the interval $[u_0 = 0,\ u_1 = a)$ by Φ^{-1}. Then there is a number $\delta > 0$ such that the mapping Φ^{-1} is defined and $1-1$ in the rectangle $u_0 \le u < u_1$, $-\delta < v < \delta$ (Fig. 19). Otherwise there would be two sequences (w_n), (w_n'), $0 \le \operatorname{Re} w_n < u_1$, $0 \le \operatorname{Re} w_n' < u_1$, $w_n \to u$, $w_n' \to u'$, with $\Phi^{-1}(w_n) = \Phi^{-1}(w_n')$ and hence $\Phi^{-1}(u) = \Phi^{-1}(u')$. It follows from the local univalence of Φ^{-1} that $u \neq u'$, and we may assume $u < u'$. Moreover, $u' - u < a$ is excluded by the minimality of $u_1 - u_0$. Therefore $u = u_0$, $u' = u_1$. But then, $w_n + a \to u_1$, $\Phi^{-1}(w_n + a) = \Phi^{-1}(w_n')$ and $w_n + a \neq w_n'$. This contradicts the local univalence of Φ^{-1} at u_1.

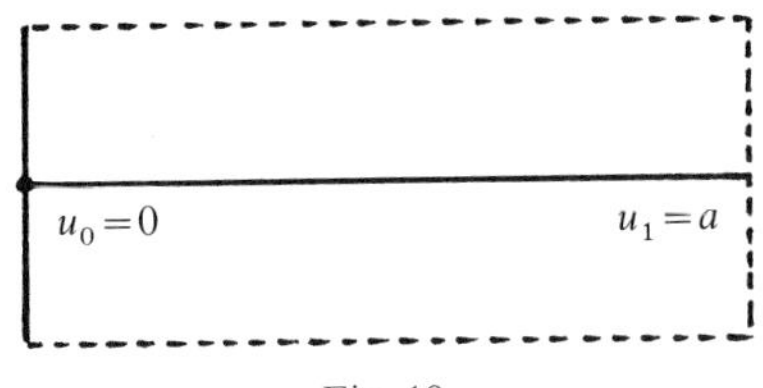

Fig. 19

The mapping Φ^{-1} is a homeomorphism of the rectangle $0 \le u \le a$, $-\delta < v < \delta$, with the points iv and $iv + a$, $|v| < \delta$, identified, onto a ring domain $R_0 \subset R$. The horizontal intervals $v = \text{const.}$ are mapped onto closed trajectories sweeping at R_0. The transformation

$$\zeta \to w = \frac{a}{2\pi i} \log \zeta$$

followed by Φ^{-1} is a conformal mapping of the circular ring

$$\exp \frac{-\delta 2\pi}{a} < |\zeta| < \exp \frac{\delta 2\pi}{a}$$

onto R_0, taking the circles $|\zeta| = \text{const.}$ into the closed trajectories in R_0. We have thus embedded the closed trajectory α_0 into a one parameter family of closed trajectories parallel to α_0.

In terms of the parameter ζ the quadratic differential φ has the representation

$$dw^2 = -\left(\frac{a}{2\pi}\right)^2 \frac{1}{\zeta^2} d\zeta^2.$$

9.4 The associated (largest) ring domain. Under the assumptions of Section 9.3 we now continue Φ^{-1} on the verticals upwards and downwards as we did it before along the real axis. For every $u \in [0, a]$ there is a maximal open interval $(\underline{v}(u), \bar{v}(u))$ on which Φ^{-1} is thus defined. The functions $\underline{v}(u)$ and $\bar{v}(u)$ are upper and lower semicontinuous respectively. For, if I is any closed subinterval of $(\underline{v}(u), \bar{v}(u))$, the conformal continuation of Φ^{-1} is defined in a neighborhood of I, by a simple compactness argument. Therefore there exist the two numbers

$$\underline{v} = \max_{0 \le u \le a} \underline{v}(u), \qquad \bar{v} = \min_{0 \le u \le a} \bar{v}(u).$$

The mapping Φ^{-1} is defined and locally $1-1$ in the rectangle $0 \le u \le a$, $\underline{v} < v < \bar{v}$ (Fig. 20). By the laws of analytic continuation, $\Phi^{-1}(iv + a) = \Phi^{-1}(iv)$ for all $v \in (\underline{v}, \bar{v})$.

Let there be two points w_1 and w_2 in $0 \le u < a$, $\underline{v} < v < \bar{v}$, with $\Phi^{-1}(w_1) = \Phi^{-1}(w_2)$. Then, as before, on the straight interval with w_1 and w_2 as its

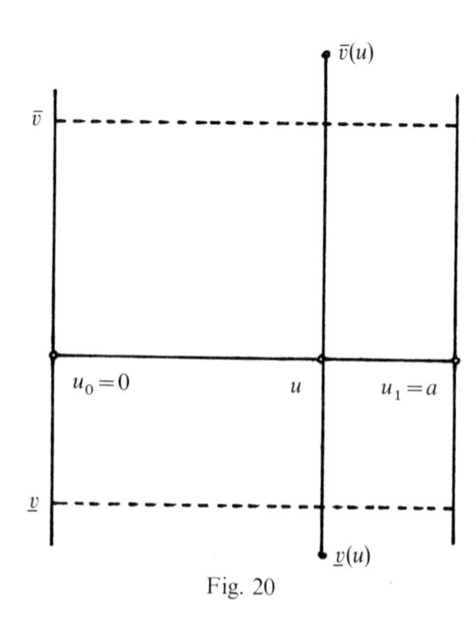

Fig. 20

endpoints, there are two such points with shortest positive distance. We call them w_1 and w_2 again. By the same reasoning as above we have $\Phi^{-1}(w+(w_2-w_1))=\Phi^{-1}(w)$ for all w near w_1. It follows that $\operatorname{Im} w_1 \neq \operatorname{Im} w_2$, because otherwise we could continue the identity for Φ^{-1} vertically to the real axis and thus get a contradiction. Now, Φ^{-1} can be continued along the line through w_1 and w_2 and obviously also horizontally, from every point of this line. It is thus defined as a doubly periodic function in the whole plane. Let w_1 be a point with smallest positive imaginary part and such that $\Phi^{-1}(w_1)=\Phi^{-1}(0)$. The parallelogram defined by the points 0, a, w_1, w_1+a is a fundamental parallelogram (Fig. 21). We have the identities $\Phi^{-1}(w)=\Phi^{-1}(w+a)$ for the

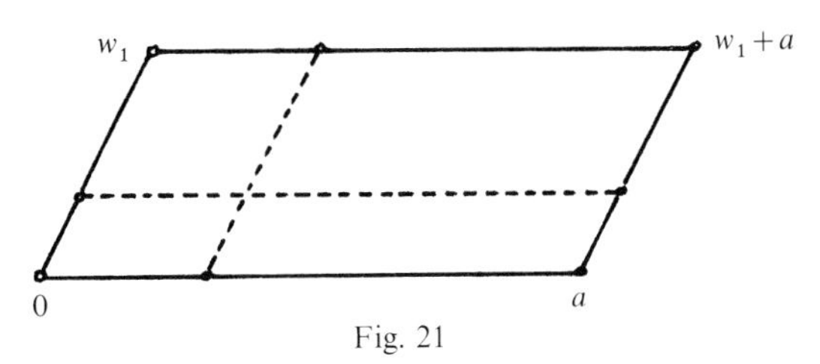

Fig. 21

points w on the left border and $\Phi^{-1}(w)=\Phi^{-1}(w+w_1)$ for the points on the lower border and there are evidently no other pairs of points in the parallelogram with the same Φ^{-1} image on R: Φ^{-1} maps the parallelogram with the proper identifications in a $1-1$ way into the surface R, and it is therefore also surjective. Thus R is a torus and φ is a holomorphic quadratic differential on R with closed trajectories.

The rectangle $0 \leq u < a$, $0 \leq v < v_1 = \operatorname{Im} w_1$ covers R exactly once, whereas $0 \leq u < a$, $0 < v < v_1$, with iv and $a+iv$ identified, represents a ring domain, swept out by closed trajectories, which covers R up to the closed trajectory $\alpha_0 = \Phi^{-1}([0, a])$.

Let us now assume that there is no such pair of points in $0 \leq u < a$, $\underline{v} < v < \bar{v}$ with the same Φ^{-1} image. As before we identify the two vertical borders of the rectangle by means of the mapping

$$w \to \zeta = \exp\left(\frac{2\pi i}{a} w\right).$$

The composition $g: \zeta \to w \to P = \Phi^{-1}(w)$ maps the annulus

$$\rho_0 = \exp\left(\frac{-2\pi}{a} \bar{v}\right) < |\zeta| < \exp\left(\frac{-2\pi}{a} \underline{v}\right) = \rho_1$$

conformally onto a region $R_0 \subset R$ containing the trajectory α_0 and swept out by closed trajectories α.

Let $\bar{v}$, say, be finite. Then there is a value u, $0 \leq u < a$, with $\bar{v}(u) = \bar{v}$. The continuation of Φ^{-1} along this vertical line is not possible above the value $\bar{v}$. If the vertical trajectory ray $\Phi^{-1}(u+iv)$, $0 \leq v < \bar{v}$, does not tend to the boundary of R, it evidently, because of its finite length, tends to a finite critical point of φ (for details see Section 10.1 etc.). Therefore the corresponding boundary curve of R_0 contains a boundary point of R or a finite critical point of φ.

The last statement of the theorem follows like this: Let $P \in R_0 \cap R_1$. Then the closed trajectory α through P is contained in both ring domains and we conclude, from the maximality, that the associated ring domain of α is equal to both R_0 and R_1.

Theorem 9.4. *Let φ be a meromorphic quadratic differential on an arbitrary Riemann surface R. Every closed trajectory α_0 of φ is embedded in a maximal ring domain R_0 swept out by closed trajectories α. It is uniquely determined except for a holomorphic quadratic differential with closed trajectories on a torus. We call it the ring domain associated with α_0. Two ring domains R_0 and R_1, associated with the closed trajectories α_0 and α_1 respectively are either disjoint or identical.*

9.5 The punctured disk and the doubly punctured sphere. We here consider the limiting cases $\rho_0 = 0$, $\rho_1 < \infty$ (equivalently $0 < \rho_0$, $\rho_1 = \infty$) and $\rho_0 = 0$, $\rho_1 = \infty$. In the first case, the composition g maps the punctured disk $0 < |\zeta| < \rho_1$ conformally (and schlicht) onto a domain $D \subset R$. According to Theorem 3.3 the point zero is either a removable singularity of g or else the point $P = g(\zeta)$ tends, for $\zeta \to 0$, to an isolated, planar and pointlike boundary component P_0 of R, which we can add to the surface. The representation of the quadratic differential in terms of the parameter ζ in D (see Section 9.3, last line) shows that P_0 is a second order pole of φ, with closed trajectories near P_0.

If $\rho_0 = 0$, $\rho_1 = \infty$, the same considerations, applied to both points $\zeta = 0$ and $\zeta = \infty$, show that R is the Riemann sphere (after addition of the two limiting points P_0 and P_∞, if necessary). The quadratic differential has the second order poles P_0 and P_∞ and all the trajectories are closed.

9.6 Freely homotopic closed trajectories. Let α_0 be a closed trajectory and R_0 its associated ring domain. Then every closed trajectory α in R_0 is evidently freely homotopic to α_0 on the surface $\dot{R} = R \setminus \{\text{poles of } \varphi\}$. Theorem 2.5 allows to prove the converse.

Theorem 9.6. *Let φ be a meromorphic quadratic differential on an arbitrary Riemann surface R. Let R_0 be the ring domain associated with a closed trajectory α_0 of φ. If a closed trajectory α_1 is freely homotopic to α_0 on the punctured surface $\dot{R} = R \setminus \{\text{poles of } \varphi\}$, then $\alpha_1 \subset R_0$, which is equivalent to saying that the ring domain R_1 associated with α_1 is equal to R_0.*

Proof. If R is the torus and φ is holomorphic on R, then, as soon as one trajectory α_0 is closed, all trajectories are closed and there is only one ring domain R_0, which covers R up to an arbitrary closed trajectory (there is no uniqueness).

In all the other cases, α_0 determines its ring domain R_0 uniquely. Let α_0 be freely homotopic to a point P on $\dot{R}$. Then it bounds a disk $D \subset \dot{R}$. Let h be a conformal mapping of D onto the unit disk $|z| < 1$ (i.e. we introduce in D the parameter z). As α_0 is an analytic Jordan curve, the function element $\varphi(z)$ is analytic in $|z| \leq 1$, and on $|z| = 1$, for tangential dz, we have $\varphi(z)\,dz^2 > 0$. Thus,

$\arg \varphi(z)\,dz^2 = \text{const.}\,(=0)$ along the circumference of the unit disk, and therefore $d\arg\varphi(z) = -2\,d\arg dz$. As the increase of $\arg dz$ along $|z|=1$ is 2π, we get

$$\frac{1}{2\pi}\int_{|z|=1} d\arg\varphi(z) = -2.$$

Therefore φ has at least one pole in D, contradicting $D \subset \dot{R}$. We conclude that α_0 is not homotopic to a point on the surface $\dot{R} = R \setminus \{\text{poles of } \varphi\}$.

Let now α_1 be freely homotopic to α_0 on $\dot{R}$. Then, the two curves bound a ring domain D on $\dot{R}$. It can be mapped conformally onto an annulus $0 < r_0 < |z| < r_1 < \infty$. The above argument, applied to this annulus, shows that φ has no zeroes in D. Therefore the ring-domain R_0 associated with α_0 contains α_1.

§10. Non Closed Trajectories

10.1 Preliminaries. Let φ be a meromorphic quadratic differential on an arbitrary Riemann surface R. In the notation of Section 9.1 let $\Phi^{-1}(\Delta) = \alpha$ be the trajectory through the point P_0, $\Phi^{-1}(0) = P_0$, $\Delta = (u_{-\infty}, u_\infty)$. We now assume that $\Phi^{-1}(u_1) \neq \Phi^{-1}(u_2)$ for all $u_1, u_2 \in \Delta$, $u_1 \neq u_2$. Then, every compact subinterval $I \subset \Delta$ is mapped homeomorphically into R and therefore is a closed Jordan arc. The length $a = |\alpha|$ of α in the φ-metric is

$$a = \int_\alpha |\varphi(z)|^{\frac{1}{2}}\,|dz| = \int_\Delta |du| = u_\infty - u_{-\infty}.$$

If we do not specify otherwise we always think of α as parametrized by u. The point P_0 which corresponds to $w=0$ as well as the positive direction on α can of course be chosen arbitrarily. The two half open subintervals into which α is subdivided by P_0 are called trajectory rays. They will be denoted by $\alpha^+ = \Phi^{-1}([0, u_\infty))$ and $\alpha^- = \Phi^{-1}([0, u_{-\infty}))$ and their orientation is supposed to be such that P_0 is the initial point of either of them.

For two arbitrarily chosen values $u_1 < u_2$ on Δ there exists a number $b > 0$ such that Φ^{-1} is a homeomorphism of the rectangle $u_1 \le u \le u_2$, $0 \le v \le b$ (respectively $-b \le v \le 0$) into the surface R. The image S is called a horizontal rectangle or rectangle adjacent to α to the left (the right) along the subinterval I bounded by the points $P_1 = \Phi^{-1}(u_1)$, $P_2 = \Phi^{-1}(u_2)$. Of course, the φ-area of S is

$$|S| = \int_0^b \int_{u_1}^{u_2} du\,dv = ab$$

with $a = u_2 - u_1$.

10.2 The limit set of a trajectory ray. (For rational quadratic differentials see Jenkins and Spencer [1].) Let $\Phi^{-1}([0, u_\infty))$ be a trajectory ray. Its limit set for $u \to u_\infty$ is denoted by A^+. It is the set of all points $P \in R$, for which there exists a sequence of numbers $u_n \to u_\infty$ such that $P_n = \Phi^{-1}(u_n) \to P$. Equivalently,

$$A^+ = \lim_{u \to u_\infty} \overline{\Phi^{-1}([u, u_\infty))}.$$

Evidently, A^+ does not depend on the initial point $P_0 \in \alpha$ of α^+.

A trajectory α has two limit sets, A^+ and A^-, according to its two rays α^+ and α^-; they are uniquely determined by α, independently of the point $P_0 \in \alpha$.

If A^+ is empty, we say that the ray α^+ tends to the boundary or is a boundary ray. A necessary and sufficient condition for α^+ to be a boundary ray is that for every compact set $C \subset R$ there is a number u_0 such that $\Phi^{-1}(u) \notin C$ for all u, $u_0 \leq u < u_\infty$. A trajectory α, both rays of which are boundary rays, is called a cross cut.

Let $P \in A^+$ be a regular point. On the trajectory γ through P, which evidently cannot be closed, we choose an arbitrary point Q. Let $I = [P, Q]$ be the closed subinterval of γ with endpoints P and Q, of φ-length $|I| = a$. Then there exists an open (schlicht) horizontal rectangle S which contains I on its middle line (Fig. 22). Let $u_n \to u_\infty$ be a sequence of real numbers such that $P_n = \Phi^{-1}(u_n) \to P$. For $P_n \in S$, the trajectory α can be continued through S. Therefore $u_\infty = \infty$, as α necessarily contains infinitely many disjoint subintervals of length a. Moreover, $u_n \pm a \to \infty$, and the points $\Phi^{-1}(u_n \pm a)$, with properly chosen signs, tend to Q. We conclude that $Q \in A^+$, and as $Q \in \gamma$ was arbitrary, $\gamma \subset A^+$.

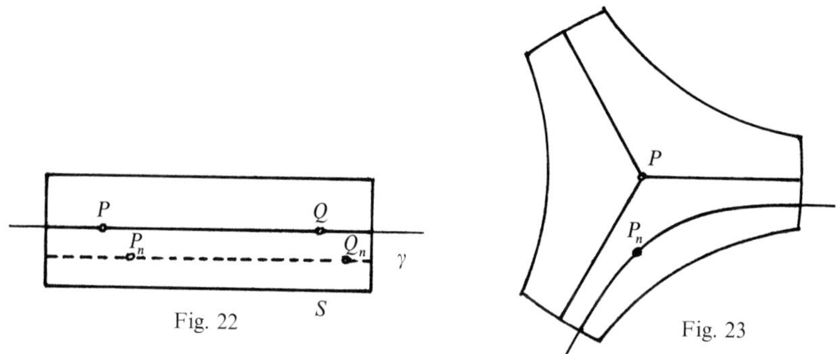

Fig. 22

Fig. 23

Let now $P \in A^+$ be a finite critical point of φ with $P_n = \Phi^{-1}(u_n) \to P$, $u_n \nearrow u_\infty$. The finitely many (see Theorem 7.1) trajectory rays ending at P subdivide a neighborhood of P into as many sectors with equal angles at P (Fig. 23). In every sector there is an adjacent closed horizontal rectangle S with P as midpoint of one of its horizontal sides. We choose them to have equal length $2a$. The interior of the union of these rectangles forms a neighborhood U of P. Unless α^+ is one of the trajectory rays ending at P, in which case $A^+ = \{P\}$, in at least one of the sectors S of U there are infinitely many points P_n and the horizontal intervals through these points (which are subintervals of α^+) extend through S. Again, $u_\infty = \infty$, $u_n \pm a \to \infty$ and the sequences of points $P_n' = \Phi^{-1}(u_n + a)$, $P_n'' = \Phi^{-1}(u_n - a)$ have points P', P'' on ∂S as accumulation points. We conclude that, if P is a zero, A^+ contains at least two neighboring rays ending at P, and if it is a first order pole, A^+ contains the ray ending at P.

Finally, if $P \in A^+$ is an infinite critical point, the ray α^+ enters every sufficiently small neighborhood of P and hence (see Sections 7.2, 7.3 and 7.4) tends to P. Again, $A^+ = \{P\}$.

A trajectory ray α^+ which tends to a finite critical point is called a critical ray. A trajectory α is critical, if at least one of its rays is critical. There are at most denumerably many critical trajectory rays.

We have proved: If $A^+=\{P\}$, P is either a finite critical point and α^+ a critical ray or else P is an infinite critical point. Also, if an infinite critical point (pole of order ≥ 2) P is contained in A^+, then $A^+=\{P\}$.

Let now A^+ contain more than one point. Then, every $P\in A^+$ is either a regular or a finite critical point, and therefore A^+ consists of non closed trajectories and, if critical, their limiting endpoints. With every regular P, the entire trajectory through P is contained in A^+.

A trajectory ray α^+ with a limit set A^+ that consists of more than one point is called divergent.

If the initial point P_0 of α^+ is contained in A^+, then $\alpha\subset A^+$. Therefore the closure $\bar{\alpha}\subset A^+$, as A^+ is closed. On the other hand, evidently $A^+\subset\bar{\alpha}$, hence $\bar{\alpha}=A^+=A$. Conversely, if $\bar{\alpha}=A^+$, clearly $P_0\in A^+$. A trajectory ray α^+ with $P_0\in A^+$ is called recurrent. A trajectory both rays of which are recurrent is called a spiral.

Let α be a trajectory with a recurrent ray α^+. Then it is easy to see by examples (see §12) that the ray α^- can be critical or a boundary ray. But it cannot tend to an infinite critical point. For, let $A^-=\{P\}$ be a pole of order ≥ 2. Choose a neighborhood U of P such that the point $P_0\in\alpha$ lies outside of U and that every ray which enters U tends to P. Now let $P_1\in a^-$, $P_1\in U$. As $P_1\in A^+$, there is a point $P_2\in\alpha^+$, $P_2\in U$. Thus α^+ enters U and must therefore tend to P, which is a contradiction.

The closure of the intersection of the limit set A^+ of a recurrent ray α^+ with an open vertical interval β is a perfect set. It is evidently relatively closed. Let $P\in A^+\cap\beta$ be a limit point of the sequence (P_n), $P_n=\Phi^{-1}(u_n)$, $u_n\to\infty$ (Fig. 24). By small shifts of the numbers u_n we can achieve that $P_n\in\beta$. But since $\alpha\subset\bar{\alpha}=A^+$, $P_n\in A^+\cap\beta$, proving the theorem.

A recurrent ray α^+ passes, through every open vertical interval β, $P_0\in\beta$, in the direction determined by α^+ at P_0. Proof: Choose β. As α^+ passes in every neighborhood of P_0, it has a first intersection P_1 with β. Assume that this intersection is in the negative direction (Fig. 25). Let S be a horizontal rectangle with the subinterval $[P_0, P_1]$ of α^+ as central line. It has its two vertical sides β_0 and β_1 on β, with midpoints P_0 and P_1 respectively. The ray α^+ passes,

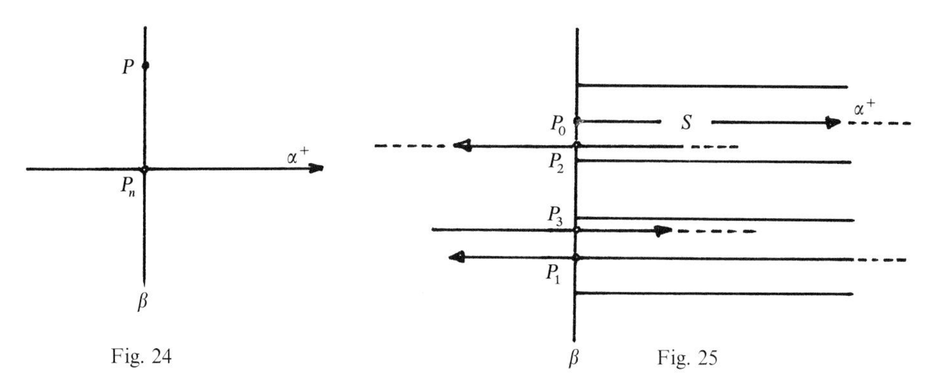

Fig. 24 Fig. 25

after the point P_1, through the interval β_0 at a point P_2, say. If this happens in the positive direction, we are done. In the opposite case we follow α^+ backwards, from P_2 and in S, all the way to the interval β_1. At the corresponding intersection point P_3, which is clearly a point of α^+, we have the desired intersection with β.

For an arbitrarily chosen point $P \in \alpha^+$ the subarc $[P_0, P]$ of α^+ intersects β at finitely many points, and a sufficiently small subinterval $\beta_0 \subset \beta$, $P_0 \in \beta_0$, has only P_0 in common with it. Therefore the point P_1 comes after P.

10.3 Half planes and horizontal strips. A trajectory ray α^+ of finite φ-length $|\alpha^+|_\varphi$, i.e. $u_\infty < \infty$, is either a critical ray or a boundary ray. If $u_\infty = \infty$, there are three possibilites: α^+ is either divergent, a boundary ray or a ray which tends to an infinite critical point.

In this section we consider non closed trajectories both rays of which have infinite length. The representation Φ^{-1} of α in terms of the natural parameter u is then defined on the entire real axis. As in Section 9.1, we continue Φ^{-1} conformally along each vertical $\operatorname{Re} w = u$, i.e. by means of the maximal open disks with centers on the vertical through u in which Φ^{-1} is schlicht. Let $(\underline{v}(u), \bar{v}(u))$ be the maximal open interval for $\operatorname{Re} w = u$ on which Φ^{-1} is thus defined. The functions $\bar{v}(u)$ and $\underline{v}(u)$ are lower and upper semicontinuous respectively. Let $\bar{v} = \inf\limits_{-\infty < u < \infty} \bar{v}(u)$, $\underline{v} = \sup\limits_{-\infty < u < \infty} \underline{v}(u)$. Of course $\bar{v} \geq 0$, $\underline{v} \leq 0$.

Assume $\bar{v} - \underline{v} > 0$, and let there be two points w_1 and w_2 in the horizontal strip $\underline{v} < v < \bar{v}$ with $\Phi^{-1}(w_1) = \Phi^{-1}(w_2)$. We may assume (see Section 9.2), that w_1 and w_2 are closest points on the line g through w_1 and w_2 with this property. Necessarily, $\operatorname{Im} w_1 \neq \operatorname{Im} w_2$, and Φ^{-1} is periodic along g with primitive period $w_2 - w_1$. By conformal continuation along the horizontals through the points of g we see that Φ^{-1} is defined and periodic, with primitive period $w_2 - w_1$, on the whole plane. If there is another pair w_3, w_4 of points, $\Phi^{-1}(w_3) = \Phi^{-1}(w_4)$, $w_4 - w_3$ not an entire multiple of $w_2 - w_1$, the function Φ^{-1} is doubly periodic and the surface R is the torus. The quadratic differential φ is holomorphic on R and the trajectories are either all closed or all non closed.

If there is no such pair w_3, w_4 we identify the point w with $w + (w_2 - w_1)$ for all $w \in \mathbb{C}$. The resulting surface is the doubly punctured sphere and Φ^{-1} is a $1-1$-conformal mapping of it into R. Therefore R itself is the doubly punctured sphere. If we represent it as the ζ-plane punctured at zero, the quadratic differential has two second order poles at $\zeta = 0$ and $\zeta = \infty$ and its trajectories are logarithmic spirals or the straight lines through zero.

There remains the case where $\Phi^{-1}(w_1) \neq \Phi^{-1}(w_2)$ for all pairs of points $w_1 \neq w_2$ in the strip $\underline{v} < v < \bar{v}$. If $\underline{v} = -\infty$, $\bar{v} = \infty$, Φ^{-1} is a $1-1$ conformal mapping of the plane into R. This surface is therefore the Riemann sphere, possibly punctured at the point which corresponds to $w = \infty$. The representation of the quadratic differential in terms of the variable w is identically equal to one. Therefore φ has a fourth order pole at infinity.

If one of the limitations is finite, the other infinite, the image $H = \Phi^{-1}(\underline{v} < v < \bar{v})$, $v = \operatorname{Im} w$, is called a (maximal) half plane with respect to φ; if both are finite, but not both equal to zero, we speak of a (maximal) horizontal strip S.

Theorem 10.3. *Two half planes, two horizontal strips or a half plane and a horizontal strip are either disjoint or else identical.*

Proof. Let D_1 and D_2 be two of the above mentioned domains. There is a 1 -1 conformal mapping Φ_1^{-1} of some S_1: $\underline{v}_1 < v < \bar{v}_1$ onto D_1 and similarly a Φ_2^{-1} of a domain S_2: $\underline{v}_2 < v < \bar{v}_2$ onto D_2. Let $P_0 \in D_1 \cap D_2$ and denote by $\zeta_i \in S_i$ the points which are mapped onto P_0 by Φ_i^{-1}, $i = i, 2$. The two function elements $w_i = \Phi_i(P)$ satisfy, in a neighborhood U of P_0, a relation of the form $\Phi_2(P) = \pm\Phi_1(P) + \text{const}$. By translations and, if necessary a change of sign in the w-planes we can achieve that $\Phi_1(P_0) = \Phi_2(P_0) = 0$ and $w = \Phi_1(P) = \Phi_2(P)$ for all $P \in U$. But then, $\Phi_1^{-1}(w) = \Phi_2^{-1}(w)$ for all $w \in U_1 = \Phi_1(U)$. Because of the maximality of the two strips S_1 and S_2 the two domains of definition of the corrected mappings Φ_1^{-1} and Φ_2^{-1} must be the same, hence $D_1 = D_2$.

10.4 Halfplanes in the neighborhood of a pole of order ≥ 3. Let P be a pole of order ≥ 3 of the quadratic differential φ, i.e. $n \leq -3$ in the notation of Sections 7.3 and 7.4. Then there are $|n| - 2$ trajectories with both rays tending to P in consecutive distinguished directions which bound, together with the point P (not necessarily maximal) half planes.

This is immediately clear if the function Φ in the neighborhood of P does not contain a logarithmic term (Section 7.3). For then, Φ has the representation, in terms of the distinguished parameter ζ,

$$w = \Phi(\zeta) = \zeta^{(n+2)/2}, \qquad |\zeta| < \rho.$$

We get $\zeta = w^k$, with $k = -\frac{2}{1}, -\frac{2}{2}, -\frac{2}{3}, \ldots, -\dfrac{2}{|n|-2}, \ldots$. The two halfplanes $\operatorname{Im} w > \rho^{(n+2)/2}$ and $\operatorname{Im} w < -\rho^{(n+2)/2}$ give rise to $|n| - 2$ φ-halfplanes of the indicated nature (see figures in Section 7.3). They can of course be enlarged to maximal φ-halfplanes on R.

If Φ has a logarithmic term, which is only possible for even n, $\sqrt{\varphi}$ has the representation, in terms of a distinguished parameter ζ,

$$dw = \left((m+1)\,\zeta^m + \frac{b}{\zeta}\right) d\zeta = \left(1 + \frac{b'}{z}\right) dz,$$

with $m = \dfrac{n}{2} = -2, -3, -4, \ldots$, $z = \zeta^{m+1}$ and $b' = \dfrac{b}{m+1}$ (see Section 7.4). Let α be the trajectory, in the z-plane, through a point z on the imaginary axis, with $\operatorname{Im} z > r$. In the domain G above α we can choose a single valued branch of the function $w = \Phi(z) = z + b' \log z$. Through every point of G there is a uniquely determined vertical trajectory. Therefore if $z' \neq z''$ are in G, then $\Phi(z') \neq \Phi(z'')$. This is evident, if they lie on the same vertical trajectory. Let them lie on two different ones, β' and β'' say. These two vertical trajectories cut α at two different points which in turn have two different Φ-images. Therefore $\operatorname{Re}\Phi(z') \neq \operatorname{Re}\Phi(z'')$. We conclude that G is mapped homeomorphically onto a Euclidean upper half plane and hence bounds a φ-half plane. A similar argument works for a trajectory through a point $z = iy$ with $y < -r$. The two half planes are taken, by the transformation $\zeta = z^{1/(m+1)}$, into $2\left(\frac{|n|}{2}-1\right)=|n|-2$ φ-half planes in the disk $|\zeta|<\rho$.

10.5 Horizontal strips

Theorem 10.5. *Let φ be a meromorphic quadratic differential on an arbitrary Riemann surface R. Then, every trajectory α, both rays of which tend to infinite critical points, is embedded in a horizontal strip.*

Proof. We first consider a ray α^+ which tends to a pole P^+ of order 2. In terms of the distinguished parameter ζ φ has the representation (see Section 7.2)

$$dw^2 = \frac{a_{-2}}{\zeta^2} d\zeta^2, \qquad |\zeta|<\rho$$

hence $w=\sqrt{a_{-2}}\log\zeta + \text{const.}$, where a_{-2} is not real and negative, as the trajectories near P^+ are not closed. We choose the constant such that the ray α^+ corresponds to a subinterval $u>u_0$ of the real axis. We conclude that, for all sufficiently large u, the mapping Φ^{-1} is schlicht on the disk of radius $\pi|a_{-2}|^{\frac{1}{2}}$ around $w=u$. This means that for all $P\in\alpha^+$ which are sufficiently close to P^+ the point P is the center of a φ-disk of radius $\pi|a_{-2}|^{\frac{1}{2}}$.

Let now P^+ be a pole of order ≥ 3. Then there is a point $P_1\in\alpha^+$ such that the vertical trajectory through P_1 bounds a half plane (i.e., in terms of the parameter w, a Euclidean half plane $\operatorname{Re} w>u_0$). Therefore Φ^{-1} is a schlicht conformal mapping of a disk $|w-u|<u-u_0$ around $w=u$. We see that $P\in\alpha^+$ is the center of an arbitrarily large φ-disk for $P\to P^+$.

Let the trajectory α be represented by Φ^{-1} on the real axis, and let the limit sets of the two rays α^+ and α^- be the infinite critical points P^+ and P^- respectively. We choose $u_1<u_2$ such that for $u\leq u_1$ and for $u\geq u_2$ the point u is the center of a disk $|w-u|<r$ on which Φ^{-1} is injective. On the other hand, the interval $[u_1,u_2]$ is contained in a rectangle $u_1\leq u\leq u_2$, $|v|\leq\delta$, on which Φ^{-1} is injective. Therefore $\bar{v}\geq\min(r,\delta)>0$ and $\underline{v}\leq\max(-r,-\delta)<0$.

§ 11. Compact Surfaces

11.1 Divergent rays on compact surfaces. The main purpose of this section is to show that every divergent trajectory ray α^+ (i.e. the limit set A^+ of which contains more than one point) is recurrent. One can say even a little bit more (Strebel [7], [8], Jenkins [3], [4]).

Theorem 11.1 *Every divergent trajectory ray α^+ of an arbitrary meromorphic quadratic differential φ on a compact Riemann surface R has the following property: Let β_0 be a closed vertical interval with the initial point P_0 of α^+ as one of its endpoints. Then, for every $P\in\alpha^+$ there is a point $P_1\in\alpha^+$ after P (i.e. $u_1=\Phi(P_1)>u=\Phi(P)$ for the appropriate branch of Φ) such that α^+ cuts β_0 at P_1 in the positive direction (i.e. in the same direction as at P_0).*

Proof. Given $P \in \alpha^+$ we may assume that β_0 is a vertical side of an adjacent rectangle of the closed interval I on α^+ with endpoints P_0 and P, that no closed trajectory passes through β_0 and that I has only P_0 in common with β_0. All this is achieved by shortening β_0, if necessary.

We mark every trajectory ray γ^+ which starts at β_0 in the positive direction and which has the following two properties:

(1) γ^+ is critical (i.e. tends to a finite critical point)
(2) γ^+ does not cross β_0 in the positive direction (except at its initial point P_0).

Consider two marked rays γ_1^+ and γ_2^+ with initial points P_1 and P_2 on β_0 respectively. If γ_2^+ were a subray of γ_1^+, the latter would cut β_0 at P_2 in the positive direction, which is excluded. Therefore each marked trajectory ray tends to a finite critical point from a different direction. Consequently, there are at most finitely many marked rays γ_i^+ and there exists a subinterval β_0' of β_0, adjacent to P_0, from which no such ray starts.

Denote by J the subinterval of α^+ with endpoints P_0 and some unspecified point $Q \in \alpha^+$. Let S be the rectangle with sides β_0' and J. As long as S is schlicht on R (i.e. Φ^{-1} is injective on the corresponding Euclidean rectangle in the w-plane) its φ-area is $|S|_\varphi = |\beta_0'|_\varphi \cdot |J|_\varphi$. If none of the trajectory rays γ^+ starting at β_0' and sweeping out S would cross β_0 again in the positive direction, we could let $Q \to \infty$ on α^+. The rectangle S would be continued indefinitely, without branching and without overlapping. But as α^+ stays outside of some neighborhoods of the infinite critical points of φ, so would S, of possibly smaller neighborhoods U_ν. We would have $|S|_\varphi \leq \iint\limits_{R \setminus \cup U_\nu} |\varphi(z)|\, dx\, dy < \infty$, which is a contradiction.

We conclude that at least one of the trajectory rays γ^+ of S must cut β_0 in the positive direction. If α^+ does, say, at $P_1 \neq P_0$ (Fig. 26), we are through, because P_1 comes after P on α^+, due to our choice of β_0. If not, we can consider a narrower rectangle S', which is bounded by an interval on α^+, an interval on α^-, a subinterval of β_0' and its mirror image on α at P_0 (Fig. 27). Proceeding with S' as before with S we find a point of α^+ on β_0 which comes after P_1 thus proving the theorem.

Corollaries. *If α^+ is a divergent trajectory ray of a meromorphic quadratic differential φ on a compact Riemann surface, then there exists a sequence of real*

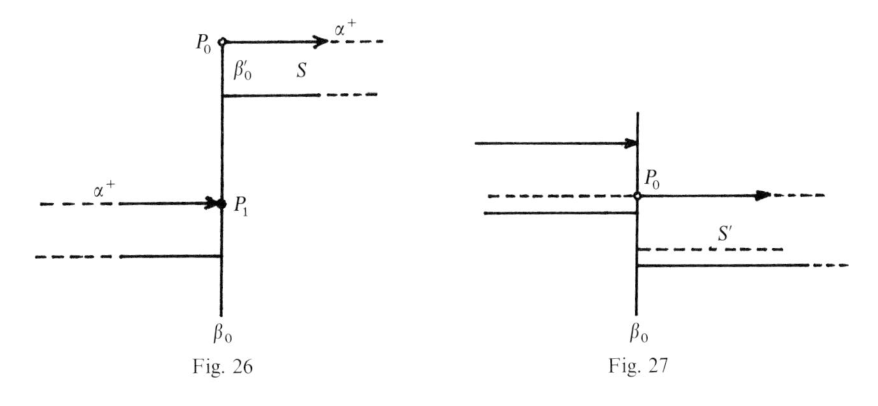

Fig. 26 Fig. 27

numbers u_n such that the points $P_n = \Phi^{-1}(u_n)$ tend to the initial point P_0 of α^+. Therefore every such ray is recurrent and its limit set A^+ is equal to the closure $A = \bar{\alpha}$ of α.

A trajectory α which carries a recurrent ray α^+ cannot, in the other direction, tend to an infinite critical point: it either tends to a finite critical point, i.e. α is a critical trajectory, or else α^- is also recurrent. We conclude that every trajectory α which carries a divergent ray is a spiral, except for finitely many critical trajectories.

Let, in the last case, the limiting critical point P be a zero. Then the two critical trajectories bordering the sectors adjacent to α at P are in A. If P is a first order pole we denote by β_1 the mirror image of β_0 with respect to α. Then, for sufficiently small β_0, there is a rectangle S adjacent to α, with vertical sides β_0 and β_1 and turning around the pole P (Fig. 28). Therefore any trajectory ray which crosses β_0 in the positive direction can be traced backwards until it meets β_1 crossing it in the negative direction. Thus α^+ will keep criss-crossing both β_0 and β_1 in opposite directions.

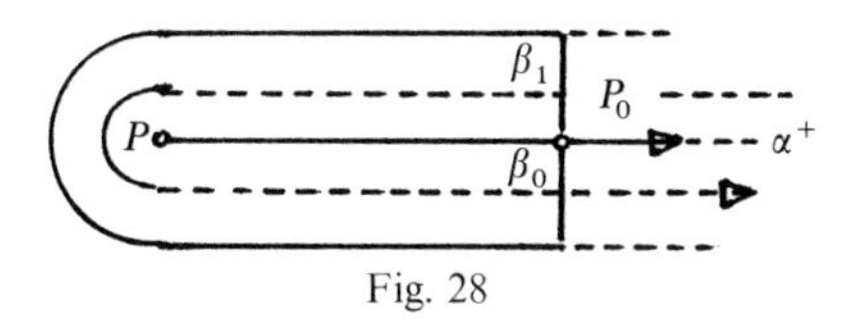

Fig. 28

11.2 The limit set of a recurrent ray. This section deals with the structure of the limit set A of a recurrent ray. It consists of regular points and of finite critical points. In the neighborhood of such a point P we can introduce a distinguished parameter and the horizontal arcs ending at P subdivide the disk $U(P)$ around P into $n+2$ congruent sectors, where $n \geq -1$ is the first exponent in the power series of φ at P. In the case of a pole we have only one sector which is the whole disk.

Theorem 11.2. *Let P be a point of the limit set A of a recurrent ray α^+. Then, if P is regular and the trajectory γ through P has infinite length, A contains a disk $U(P)$; if γ has finite length, A contains at least a half disk centered at P. Finally, if P is a critical point, there exists at least a sector of a disk around P which is in A.*

Proof. (i) Let $P \in A$ be a regular point and let the trajectory γ through P have infinite length. Choose a vertical arc β with P as one of its endpoints. As $\bar{\gamma} \subset A$, γ must carry a recurrent ray and therefore meet β at a point $P_0 \neq P$. If the interval $[P, P_0]$ on β is contained in A, a rectangle adjacent to γ and containing $[P, P_0]$ is contained in A, and so is therefore half a disk around P (Fig. 29).

If $[P, P_0] \not\subset A$, there exists a point $Q_1 \in [P, P_0]$, $Q_1 \notin A$. As A is closed, Q_1 is contained in a largest open subinterval I_1 of β which is not in A (Fig. 30). Let P_1 be the endpoint of I_1 which is situated between P and Q_1. Then $P_1 \neq P$, since γ crosses β between P and Q_1 and $\gamma \subset A$. Also P_1 is a point of some trajectory

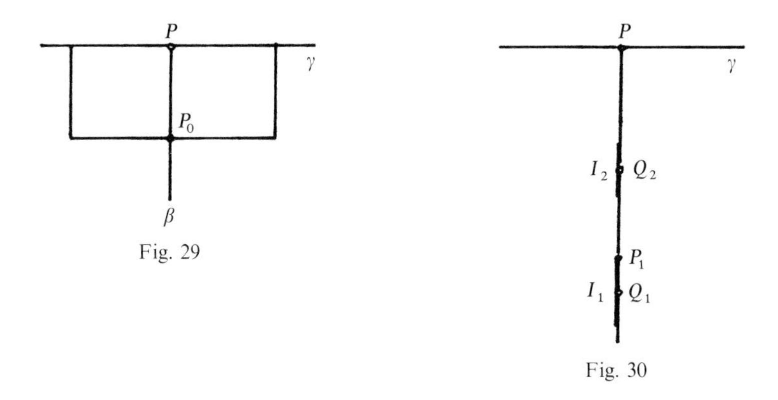

Fig. 29

Fig. 30

γ_1 and $|\gamma_1|<\infty$, since if $|\gamma_1|=\infty$ it would pass through β between P_1 and Q_1 and of course $\gamma_1\subset A$. Now if $[P,P_1]\subset A$, we are through. So assume $[P,P_1]\not\subset A$. Then there exists a point $Q_2\in[P,P_1]$, $Q_2\notin A$. Again we have a largest open interval $I_2\subset\beta$, $Q_2\in I_2$, which has no point in common with A. We consider its endpoint $P_2\in[P,Q_2]$ and repeat for P_2 the argument applied to P_1. We get a critical γ_2, $P_2\in\gamma_2$ and $|\gamma_2|<\infty$. It is not excluded that $\gamma_2=\gamma_1$, but as each of the trajectories γ_i passes through a fixed rectangular neighborhood of P and the total length of all critical trajectories of finite length is bounded, this process must stop after a finite number of repetitions. Therefore there exists a point $P_n\in\beta$, $P_n\neq P$, with $[P,P_n]\subset A$. A similar reasoning applied to the vertical interval β' which is symmetric to β with respect to γ completes the proof.

(ii) Assume that $P\in A$ is regular and lies on a trajectory γ of finite length. The trajectory ray α^+, say, must cross arbitrarily close to P on at least one side of γ. Choosing β as above, on the proper side of γ, α^+ intersects β in every neighborhood of P. This enables us to make the same construction but using α instead of γ. It follows that at least one half neighborhood of P is contained in A.

(iii) If P is a finite critical point, then in at least one of the sectors determined by the critical trajectories ending at P the ray α^+ must pass arbitrarily close to P. The above argument, applied to a vertical arc in this sector shows that its intersection with a sufficiently small disk $U(P)$ lies in A.

Corollaries. (1) *The interior A^0 of the limit set of a recurrent ray α^+ is not empty and connected, hence a domain.*

Proof. Every point $P\in\alpha$ is regular, in A, and as $|\alpha|_\varphi=\infty$ there is a neighborhood U of P in A. Hence $\alpha\subset A^0$. If P' is an arbitrary point of A^0, the ray α^+ meets every neighborhood U' of P'. Therefore U' can be connected with the initial point P_0 of α^+ by a subarc of α^+, which is in A^0.

(2) *The boundary of A consists of critical trajectories of finite length (i.e. both rays of which are critical) and their limiting finite critical points.*

Proof. Let $P\in\partial A$. Then if P is regular, the trajectory γ through P must have finite length and lies entirely in A, as well as its limiting critical points. If P is critical it follows from (iii) that the intersection of a disk U around P with A consists of entire sectors of U. At least one of the sectors has no interior point in common with A. (Therefore P must be a zero; the first order poles of A are in A^0.) Of course every critical trajectory which separates two sectors, one in A, the other outside, is entirely in ∂A.

(3) *Every trajectory ray α_1^+ of infinite length passing through a point of A is everywhere dense in A, and its limit set A_1 is equal to A.*

Proof. As $\alpha_1 \subset A$, α_1^+ must be recurrent. Let $P_1 \in \alpha_1$. There is a neighborhood $U(P_1) \subset A_1$ and thus a point $P \in \alpha$, $P \in A_1$. We conclude $\alpha \subset A_1$, $\bar{\alpha} = A \subset A_1$. Conversely, $\alpha_1 \subset A$, hence $\bar{\alpha}_1 = A_1 \subset A$, showing $A_1 = A$. Since α_1 is dense in A_1, it is dense in A.

(4) *If the limit sets of two recurrent rays have an interior point in common, they coincide.*

Proof. Let $P \in A_1^0 \cap A_2^0$. Then there is a point P_1 of the generating ray α_1^+ in A_2, hence $A_1 = A_2$ according to corollary (3).

11.3 Partitioning of A into horizontal rectangles (Strebel [8], [9]). A ring domain swept out by closed trajectories can be cut along a vertical cross section and in the resulting simply connected domain we can introduce the variable $w = \Phi(P)$. Something similar will now be done for the limit set A of a recurrent ray, but it is of course more complicated and besides, there is much more freedom.

Let A be the limit set of a recurrent ray. We choose a closed vertical interval β in A and orient it so that we can speak of a right edge β^+ and a left edge β^- (we rather consider β as a slit). On β^+ we mark, besides the two endpoints, the initial points of those rays γ^+ which end in a critical point or in one of the endpoints of β before crossing β in any direction. The same thing we do for the left edge β^-, but this time with respect to the rays γ^- starting on β^-. The finitely many open intervals between the markings on β^+ are called I_μ^+, the ones between the markings on β^- are called I_ν^-. Let now α^+ be a trajectory ray starting at a point $P^+ \in I_\mu^+$. Denote its first intersection with β by Q^- or Q^+, depending on whether it is on the left edge β^- or on the right edge β^+ of β (i.e. whether α^+ tends to this point from the left or from the right side). Q (i.e. Q^- or Q^+) cannot be a marked point on β, therefore it is an interior point of some interval I_λ (either an I_λ^- or an I_λ^+). There exists an open rectangle S with its vertical sides on β and containing the interval $[P^+, Q]$ as a diameter (Fig. 31). Every compact subinterval of I^+ can be covered by the initial vertical intervals of finitely many such rectangles, necessarily overlapping, hence of the same length. Their union is again such a horizontal rectangle, with the two vertical sides in I_μ^+ and I_λ respectively. We conclude that there is a rectangular strip S_μ with I_μ^+ as initial vertical interval and with terminal vertical interval in I_λ; reversing the argument shows that it is actually equal to I_λ.

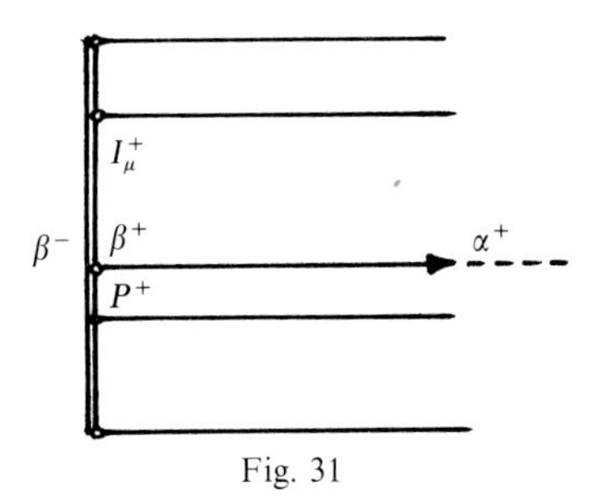

Fig. 31

Analogously we construct the rectangular strips for the intervals I_ν^-. The resulting strips S are of two different kinds: The strips of the first kind have their two vertical sides on the two different edges of β, the strips of the second kind on the same edge, either both on β^+ or both on β^-. Their totality is called the horizontal strips to the base β.

We claim that the union $\bigcup \bar{S}$ of the closed horizontal strips to the base β is equal to A, for every closed vertical interval $\beta \subset A$.

As $\beta \subset A$, every trajectory through a point of β is in A, so $\bar{S} \subset A$ for every S. To prove the converse, let α^+ be a recurrent ray which generates A (i.e. $\bar{\alpha}^+ = A$) and which does not go through an endpoint of β. We may assume that its initial point P_0 lies on β, as α^+ must cut β and we are free to choose P_0 on α^+. Let $P \in \alpha^+$, $P \notin \beta$ (for $P \in \beta$ there is nothing to prove). Denote its next intersection with β, after P, with Q. If it is a point Q^- (i.e. if α^+ approaches β from the left at Q), then it lies in one of the open intervals I_ν^- of β^-, hence P lies in the corresponding horizontal strip S, and likewise if it is a point Q^+. As $P \in \alpha^+$ was arbitrary, $\alpha^+ \subset \bigcup \bar{S}$, therefore $\bar{\alpha}^+ = A \subset \bigcup \bar{S}$.

11.4 The global trajectory structure on a compact surface (see also Jenkins [3], [4]). The trajectories of a meromorphic quadratic differential φ on a compact surface are of the following types:

1) Closed trajectories (see §9). They sweep out disjoint ring domains, including the possibility of punctured disks, the punctures corresponding to second order poles (Section 7.2). The two ring domains R_1 and R_2 determined by the closed trajectories α_1 and α_2 coincide if and only if α_1 and α_2 are freely homotopic on the surface $\dot{R} = R \setminus \{\text{poles of } \varphi\}$ (see Section 9.2). The ring domain R_0 associated with a closed trajectory α_0 has two degenerated boundary components iff R is the Riemann sphere and φ has two second order poles with negative real coefficient a_{-2} (Section 7.2 and 9.5).

2) Non closed trajectories.

a) Critical trajectories. At least one ray, say α^+, of such a trajectory α tends to a finite critical point. The other ray either tends to another finite critical point (possibly the same), to an infinite critical point or else it is recurrent. The length $|\alpha|_\varphi$ is finite iff α^+ and α^- tend to finite critical points. There are only finitely many critical trajectories.

b) Trajectories both rays of which tend to infinite critical points. The representation Φ^{-1} of α is defined on $\Delta = \mathbb{R}$. As shown in Section 10.5, $\underline{v} < 0$, $\bar{v} > 0$. Let $\underline{v} = -\infty$, $\bar{v} = \infty$. Then (see Section 10.4), either Φ^{-1} is a $1-1$

conformal mapping of the plane into the surface R, or it is simply periodic. In the first case, R is the Riemann sphere and φ has a single pole of order four at $w=\infty$, the representation of φ in terms of the variable w being $\varphi(w)\equiv 1$. In the second case, R is again the Riemann sphere and Φ^{-1} is a conformal homeomorphism of the plane, after identification of equivalent points, into R. The quadratic differential φ has two second order poles (with non closed trajectory structure).

If, say, $\underline{v}>-\infty$, $\bar{v}=\infty$, Φ^{-1} is schlicht in $v>\underline{v}$. We have a φ-half plane on R. The point $w=\infty$ clearly corresponds to a pole of order ≥ 3.

Finally, if $-\infty<\underline{v}<\bar{v}<\infty$, the trajectory α is embedded into a horizontal strip. The boundary consists of critical trajectories with no recurrent rays.

c) Spirals, i.e. trajectories both rays of which are recurrent. The interior A^0 of the closure $A=\bar{\alpha}$ of α is a domain bounded by critical trajectories of finite length.

Since these are the only possibilities and no two of the listed domains can have points in common unless they coincide, the surface R minus the critical points and the critical trajectories with no recurrent rays is subdivided into ring domains (including punctured disks), half planes, horizontal strips and spiral domains (interior of the closure of a spiral). (For a topological study of the trajectory structure see the papers of W. Kaplan [1], [2], [3] and Jenkins [5].)

§ 12. Examples

12.1 Holomorphic quadratic differentials on the torus (see also Section 4.5). Let φ be a holomorphic quadratic differential on a torus R. The surface R can be represented as a parallelogram with the corners 0, 1, ω, $1+\omega$ ($\operatorname{Im}\omega>0$) and with the identifications $z\leftrightarrow z+1$, $z\leftrightarrow z+\omega$ on the sides (Fig. 32). φ is holomorphic on the closed parallelogram, and in the neighborhood of the points z_0, z_0+1 we have $\varphi(z)\,dz^2=\varphi(z+1)\,d(z+1)^2=\varphi(z+1)\,dz^2$. Therefore $\varphi(z)=\varphi(z+1)$, and likewise $\varphi(z)=\varphi(z+\omega)$ on the boundary. The function $\varphi(z)$ can therefore be continued as a holomorphic and doubly periodic function on the complex plane, hence $\varphi=c=\text{const}$. The trajectories are the straight lines $c\,dz^2>0$, i.e. $\arg dz=-\frac{1}{2}\arg c \pmod{\pi}$. Let α be the trajectory through zero and assume that it cuts the side $[1, 1+\omega]$ at some point z. Then, as is easily seen, α is closed on

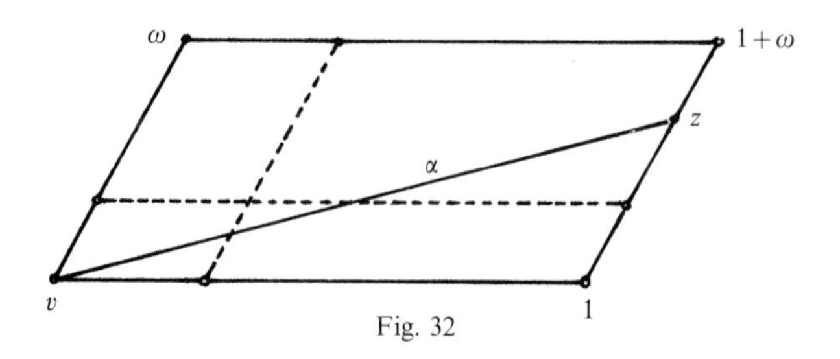

Fig. 32

R if and only if the interval $[1, z]$ is a rational multiple of $[1, 1+\omega]$. If the quotient of the two intervals is irrational, both rays of α are everywhere dense on R, even equally distributed. This means the following: Let $P_0 = 0$ and let P be an arbitrary point of α^+. Denote the number of intersections of the segment $[P_0, P]$ of α^+ with an arbitrary fixed subinterval I of the side $[1, 1+\omega]$ by n_I, whereas n is the total number of intersections of $[P_0, P]$ with the side $[1, 1+\omega]$. Then, with the length of $[P_0, P]$ tending to infinity, the quotient $n_I : n$ tends to the quotient of the lengths of I and $[1, 1+\omega]$. Another way to put it: Let β_1 and β_2 be two arbitrary vertical intervals on R, with φ-lengths b_1 and b_2 respectively (the φ-length, in this case, is $|c|^{\frac{1}{2}}$ times the Euclidean length in the z-plane). Then the numbers of intersections n_{β_1} and n_{β_2} have the property that their quotient tends to $b_1 : b_2$.

It should be noted that φ is the square of the linear differential $\sqrt{c}\,dz$.

12.2 Meromorphic quadratic differentials of finite norm on the Riemann sphere. Let φ be a meromorphic quadratic differential, not identically zero, on the Riemann sphere $R = \hat{\mathbb{C}}$. The surface R can be covered by two neighborhoods U: $z \neq \infty$ and $\tilde{U}$: $\tilde{z} = \frac{1}{z} \neq \infty$. Let the corresponding meromorphic function elements be $\varphi(z)$ and $\tilde{\varphi}(\tilde{z}) = \tilde{z}^n (a_n + a_{n+1}\tilde{z} + \dots)$ respectively, $a_n \neq 0$. For $z \to \infty$ we have, because of the transformation rule $\varphi(z)\,dz^2 = \tilde{\varphi}(\tilde{z})\,d\tilde{z}^2$, $\varphi(z) = \frac{1}{z^{n+4}}\left(a_n + \frac{a_{n+1}}{z} + \dots\right)$. Therefore, $\varphi(z)$ is a rational function. Conversely, if we take any rational function $\varphi(z)$, we can consider it as a quadratic differential on the extended plane $\hat{\mathbb{C}}$. By the transformation rule, the order at infinity is lowered by four. Therefore the sum of the orders of the zeroes minus the sum of the orders of the poles of $\varphi(z)$, if considered as a quadratic differential, is minus four.

If φ has to have finite norm $\|\varphi\| = \iint_{\mathbb{C}} |\varphi(z)|\,dx\,dy$, then it can only have poles of the first order. Therefore, the simplest quadratic differentials of finite norm on the Riemann sphere are the ones with four simple poles. If they are finite, $z_1, \dots, z_4$ say, the function $\varphi(z)$ has to disappear with at least the fourth order for $z \to \infty$. Therefore, the quadratic differentials of finite norm on the extended plane $\hat{\mathbb{C}}$ with exactly the poles $z_1, \dots, z_4 \in \mathbb{C}$ are of the form

$$\varphi(z) = \frac{c}{(z-z_1)\dots(z-z_4)}$$

with arbitrary complex constant c. If $z_4 = \infty$, we have

$$\varphi(z) = \frac{c}{(z-z_1)(z-z_2)(z-z_3)},$$

which only disappears with the third order for $z \to \infty$. This means that, in terms of the variable $\tilde{z} = \frac{1}{z}$, the point $\tilde{z} = 0$ becomes a first order pole.

In order to get an insight into the trajectory structure of φ we lift it to the two sheeted covering surface $\tilde{R}$ of R with branchings exactly over the four points $z_1, \ldots, z_4$. We get it by cutting the plane along two non intersecting arcs γ_1 and γ_2 joining z_1 to z_2 and z_3 to z_4 respectively and then glueing two identical copies of the cut plane by crosswise identification along the two cuts. The lifting is performed by means of the projection map π. In the neighborhood of a branching point z_k we introduce on $\tilde{R}$ the parameter $\tilde{z}=\sqrt{z-z_k}$. Therefore $\pi(\tilde{z})=z-z_k=\tilde{z}^2$ is the projection map in the neighborhood of the point $\tilde{P}_k$ above z_k. Let the representation of φ, in terms of the variable $z-z_k$ near z_k be

$$\varphi(z)=(z-z_k)^n(a_n+a_{n+1}(z-z_k)+\ldots).$$

Then, using the transformation rule $\tilde{\varphi}(\tilde{z})\,d\tilde{z}^2=\varphi(z)\,dz^2$, we get for the lift $\tilde{\varphi}$ in terms of $\tilde{z}$ near $\tilde{P}_k$

$$\tilde{\varphi}(\tilde{z})=\tilde{z}^{2n}(a_n+a_{n+1}\tilde{z}^2+\ldots)(2\tilde{z})^2=\tilde{z}^{2n+2}(4a_n+4a_{n+1}\tilde{z}^2+\ldots), \qquad a_n\neq 0.$$

In particular, if $n=2m+1$ is odd, we have

$$\tilde{\varphi}(\tilde{z})=\tilde{z}^{4m+2}(4a_{2m+1}+4a_{2m+2}\tilde{z}^2+\ldots),$$

which is equal to $4a_{-1}+4a_0\tilde{z}^2+\ldots$ for $m=-1$. Therefore, the lifted quadratic differential $\tilde{\varphi}$ is regular (i.e. different from zero) at the points $\tilde{P}_k$ and thus has no critical points on R, which is a torus.

In the neighborhood of the two points of R above a point $z_0\neq z_k$ we can use the local parameter $\tilde{z}=z-z_0$. In terms of this variable, we have $\tilde{\varphi}(\tilde{z})=\varphi(z)$ and $d\tilde{z}=dz$. Therefore the trajectories are the same on both sheets and they project onto the trajectories of φ in the plane. From every first order pole z_k emerges a critical trajectory α of φ; its lift $\tilde{\alpha}$ consists of two branches reflected at $\tilde{P}_k$ but lying in the two different sheets of $\tilde{R}$. The trajectory $\tilde{\alpha}$ evidently is closed on $\tilde{R}$ if and only if it goes through another branching point $\tilde{P}_j$. Therefore φ has closed trajectories iff it has two critical trajectories joining pairs of first order poles. The two arcs are the two boundary components of a ring domain swept out by closed trajectories.

In the other case (non closed trajectories) there are four critical trajectories consisting of a critical ray (which ends in one of the critical points) and a recurrent ray. As on the torus, every ray, except for the four critical ones, is everywhere dense in the plane, even equally distributed (in the sense of Section 12.1).

12.3 Welding of surfaces. In order to get more examples of different trajectory structures we perform a sewing process. This process allows a welding of two surfaces with given quadratic differentials to a single surface R and a quadratic differential φ on R.

Let R_1 and R_2 be arbitrary Riemann surfaces with given quadratic differentials φ_1 and φ_2 respectively. We choose two closed horizontal arcs γ_1 and γ_2 of φ_1 and φ_2 respectively, of the same length $|\gamma_1|_{\varphi_1}=|\gamma_2|_{\varphi_2}$. Since φ_2 and $\lambda\varphi_2$ have the same trajectories for any positive number λ, this may be assumed without any loss of generality.

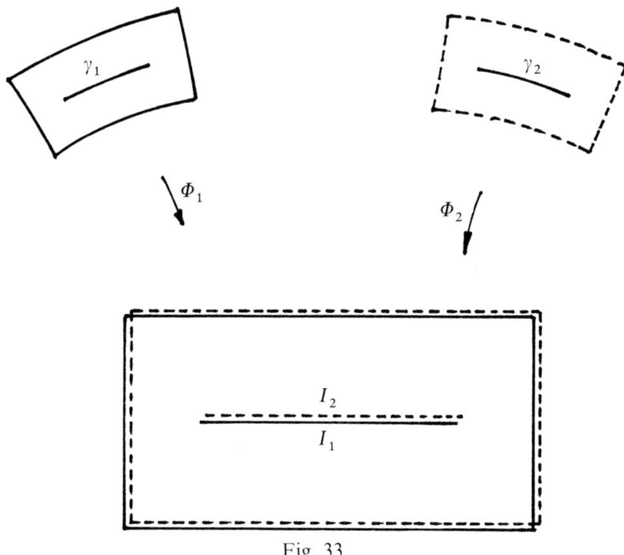

Fig. 33

We now enclose the two arcs in open horizontal rectangles S_1 and S_2, which are then mapped by Φ_1 and Φ_2 respectively onto open Euclidean rectangles in the w-plane in such a way that the two images $I_1 = \Phi_1(\gamma_1)$ and $I_2 = \Phi_2(\gamma_2)$ coincide (Fig. 33). We cut the two rectangles $S'_k = \Phi_k(S_k)$ along the intervals I_k and connect them crosswise to form a new two sheeted piece of surface. In terms of the variable w in S'_k both quadratic differentials have the representation $\varphi_k \equiv 1$. Therefore they are analytic continuations of each other across the slits and thus form together a single quadratic differential φ on the surface R which is the union of R_1 and R_2 by crosswise identification of the slits I_1 and I_2. In the neighborhood of the two identified initial and endpoints of I_1 and I_2 we have to introduce the square root as a parameter on R. There are four critical trajectories emanating from each of these points.

In this way we can e.g. combine two tori R_1 and R_2 to a single surface R of genus 2. If we start with holomorphic quadratic differentials φ_1 and φ_2, we get again a holomorphic quadratic differential φ with two zeroes of the second order. The trajectory structure is just taken over from φ_1 and φ_2 in each of the respective parts.

If we apply the procedure to two quadratic differentials with four first order poles each on the Riemann sphere, we get a quadratic differential φ with eight first order poles and two second order zeroes (therefore of finite norm) on a surface R which is again the Riemann sphere. Therefore φ is hyperelliptic, and the trajectory structure can be read off from the original ones.

Instead of starting with two horizontal arcs we can start with vertical arcs γ_1 and γ_2. The mappings Φ_1 and Φ_2 take them into vertical intervals I_1 and I_2 of the w-plane and the representation of φ_1 and φ_2, in terms of the variable w is again identically equal to one, valid in a neighborhood of the slits. Therefore, crosswise identification and analytic continuation is again possible. But this time the two trajectory structures will mix, because they cross over the closed curve, consisting of two critical vertical trajectories, along which R_1 and R_2 are welded.

12.4 Interval exchange transformations. We consider a meromorphic linear differential φ on a compact surface R. The requirement $\varphi(z)\,dz>0$ defines a field of half lines on R, and of course everything which has been said about the trajectories of a quadratic differential is true for the trajectories of φ; besides, they can be oriented globally. In particular, every divergent ray α^+ is recurrent, and every vertical interval β in its limit set A^+ is the basis of a set of horizontal rectangles S which cover A^+ (see Section 11.3). But now, only strips S of the first kind are present. Their initial vertical intervals I_μ^+ lie on the positive border β^+ of β, whereas the terminal vertical intervals I_μ^- are subintervals of β^-. Of course, $|I_\mu^-|_\varphi=|I_\mu^+|_\varphi$; the intervals I_μ^- on β^- are simply a permutation of the intervals I_μ^+ on β^+. We now define a transformation T of β onto itself: For $x\in I_\mu^+$ the image Tx is the point in I_μ^- which results from the translation of I_μ^+ onto I_μ^-. In order to make T single valued, we choose the intervals I_μ^+ and I_μ^- as well as β half open to the right: that means that T is, at the endpoints of the intervals I_μ^+, continuous from the right. Of course, by the mapping Φ, the arc β can be considered as a straight vertical interval I and the length as Euclidean length. Such a transformation T is called *interval exchange transformation* on I (Strebel [8], Keane [1], Keynes and Newton [1], Masur [3]). We have thus seen that a linear differential φ on a compact Riemann surface with a recurrent trajectory ray α^+ defines an interval exchange transformation on every vertical interval β which lies in the limit set $\bar{\alpha}$ of α^+. Roughly speaking we assign to every $P\in\beta$ the next intersection of the ray $\alpha^+(P)$ which starts at P with the arc β.

Conversely, let I be an arbitrary half open vertical interval, partitioned into half open intervals I_ν^+, $\nu=1,\dots,p$, and let π be a permutation of the numbers $1,\dots,p$. I is now partitioned into the intervals $I_1^-,\dots,I_p^-$, with $|I_k^-|=|I_{\pi(k)}^+|$. This defines an interval exchange transformation T which translates I_μ^+ to $I_{\pi^{-1}(\mu)}^-$. We construct a (bordered) Riemann surface R and a linear differential φ on R such that its trajectory structure (which can have closed trajectories) defines T on I. To this end,

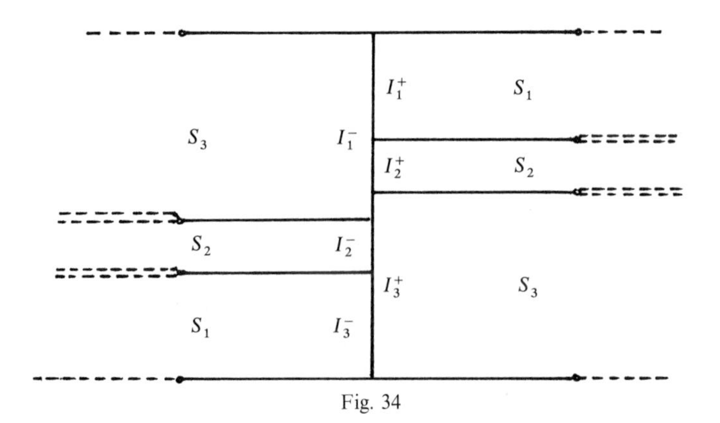

Fig. 34

we attach horizontal rectangles S_ν to the intervals I_ν^+, of length one, say. The terminal vertical interval of S_ν is identified with $I^-_{\pi^{-1}(\nu)}$, and the horizontal sides of two neighboring rectangles (on their initial resp. terminal vertical side) are identified from I to the left and to the right, up to a distance of one third from I (Fig. 34). This is a model of a bordered Riemann surface R. As local parameter in the neighborhood of an interior point or a smooth boundary point P_0 of R we use $\tilde{z}=z-z_0$, where z_0 is an underlying point of the complex plane (it is uniquely determined for every P_0 up to a translation). At the bifurcation points z_k, we use $\tilde{z}=\sqrt{z-z_k}$ as a local homeomorphism. It maps a neighborhood of the corresponding point of R onto a half disk. The linear differential φ has, in terms of the local parameter z, the representation $\varphi(z)\equiv 1$; at the bifurcation points P_k we get, by the transformation rule $1\cdot dz=\tilde{\varphi}(\tilde{z})\,d\tilde{z}$, the representation $\tilde{\varphi}(\tilde{z})=2\tilde{z}$. The points P_k are first order zeroes of φ, and second order zeroes of the corresponding quadratic differential φ^2 on R.

In order to get a compact Riemann surface, one can take the double of R and continue φ by reflection. We have thus shown: Every interval exchange transformation is generated by the trajectory structure of a holomorphic linear differential on a compact Riemann surface.

As a specific example we consider a holomorphic linear differential φ with nonclosed trajectories on a torus R and a closed vertical interval β on R. In accordance with Section 11.3 we mark, on the positive edge β^+, the two endpoints of β and the initial points of the rays which go through the endpoints of β before they traverse β. The following two situations shown in Figs. 35 and 36 are possible:

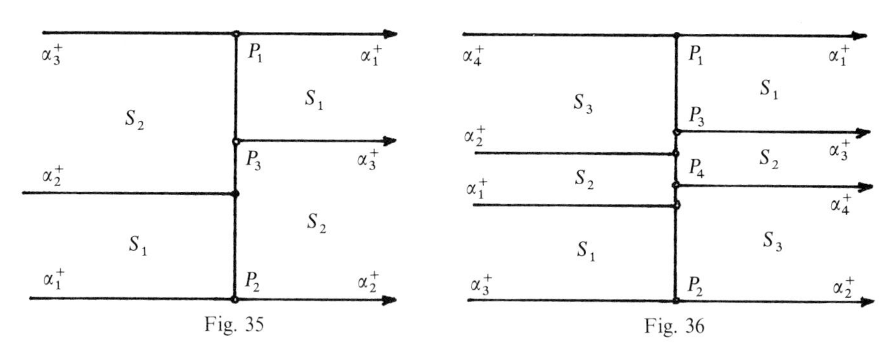

Fig. 35 Fig. 36

Denote the two endpoints of β with P_1 and P_2, and the rays emanating from these points with α_1^+ and α_2^+ respectively. In the first case, one of these rays, α_1^+ say, has as next point after P_1 the point P_2 in common with β (i.e. α_2^+ is a subray of α_1^+). Then, there is one other marked point P_3, with its ray α_3^+ passing through P_1. We have thus two intervals $I_1^+=[P_1,P_3]$ and $I_2^+=[P_3,P_2]$, with corresponding strips S_1 and S_2. Their end intervals must be interchanged (otherwise the trajectories would be closed), and the transformation T is defined by the transposition of two intervals.

In the second case, there are four marked points on β^+, determining three intervals $I_1^+=[P_1,P_3]$, $I_2^+=[P_3,P_4]$, $I_3^+=[P_4,P_2]$, which are the initial vertical sides of the strips S_1, S_2 and S_3. It is easy to see that the terminal vertical sides of the strips are in the order 3, 2, 1, because α_3^+ must go to P_2 and α_4^+ to P_1. Therefore T is determined by three intervals which are permuted in the order $(1, 2, 3) \to (3, 2, 1)$. In the first case T operates on I as a translation $\operatorname{mod}(|I|)$ by $|I_2^+|$, whereas, in the second case, it can be considered as a translation by $|I_2^+| + |I_3^+| \operatorname{mod}(|I| + |I_2^+|)$.

Conversely, every interval exchange transformation with two or three intervals which is of the above form is induced by the trajectories of a holomorphic linear differential on a torus. In the first case, this is easy to see: The torus R is represented by the square $0 \le x \le 1$, $0 \le y \le 1$, with identified opposite sides, and the trajectories are as in Fig. 37. T operates on the closed line $0 \le x \le 1$, $y = 0$ on R.

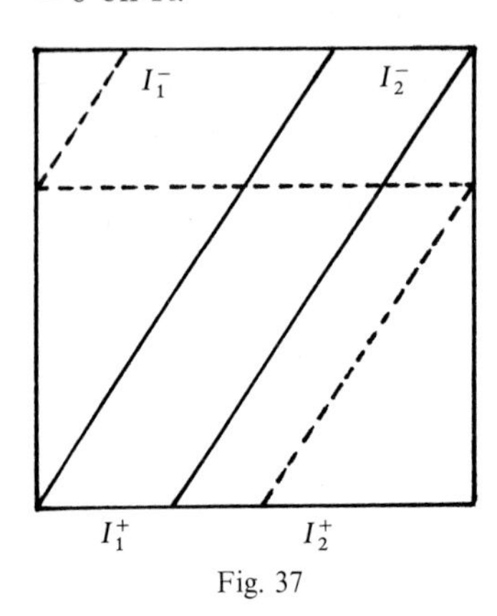

Fig. 37

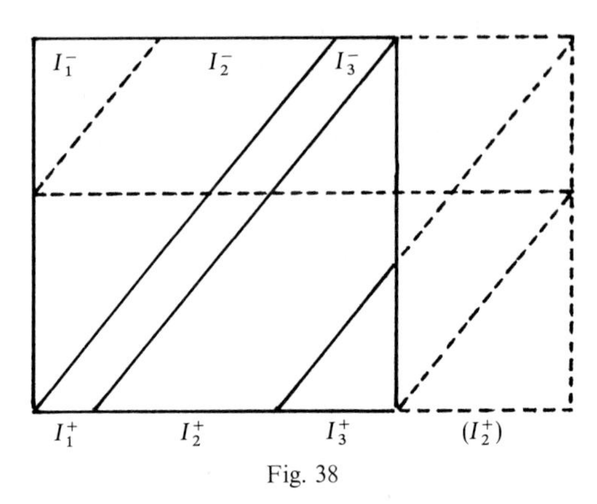

Fig. 38

In the second case, the torus is represented by the square, with base I, plus a rectangle of width $|I_2^+|$ (Fig. 38). T operates on I again, but this time I is not closed on R.

We conclude that for an interval exchange transformation of this kind the orbits $T^n(P)$ are either closed, for all points $P \in I$, or equally distributed on I. This is no longer the case for interval exchange transformations with more than three intervals (see Keynes and Newton [1]).

§ 13. Quadratic Differentials with Finite Norm

In this § the trajectory structure of a holomorphic quadratic differential of finite norm on an arbitrary Riemann surface is studied. It is shown that there are, except for a set of measure zero, only three simple types of trajectories. This leads to a partitioning of the surface with respect to a given quadratic differential into subsets covered by the different kinds of trajectories, a procedure which is used in connection with the length area method.

Only holomorphic quadratic differentials are considered. The case of a meromorphic quadratic differential of finite norm is reduced to the former one by puncturing the surface at the first order poles. At each pole there is a critical trajectory emanating from it.

The case of a simply connected domain (the disk) is treated independently and in much more detail in §19.

13.1 Exceptional trajectories (Strebel [13]). A trajectory ray α^+ of a holomorphic quadratic differential φ on an arbitrary Riemann surface R can have one of the following characteristic behaviours (see Section 10.2):

a) Its limit set A^+ is empty. The ray α^+ tends to the boundary and is therefore called a boundary ray. A trajectory both rays of which tend to the boundary is called a cross cut of R.

b) Its limit set A^+ consists of a single point P. Then P is a zero of φ and α^+ tends to P in one of the distinguished directions. Such a ray is called critical; a trajectory with at least one critical ray is called critical itself. There are at most denumerably many critical trajectories.

c) A^+ consists of more than one point. In this case, α^+ is called divergent; A^+ is a union of trajectories and, if critical, their limiting critical points on R. If, in particular, the initial point P of α^+ is in A^+, we have $A^+=\bar{\alpha}$. The ray α^+ is said to be recurrent. A trajectory α with two recurrent rays is called a spiral.

Definition 13.1. A trajectory of a holomorphic quadratic differential of finite norm, which is neither closed, nor a cross cut of finite length, nor a spiral is called exceptional.

An exceptional trajectory α of a holomorphic quadratic differential of finite norm can have one of the following behaviours (which are not mutually exclusive):

(1) α is critical,

(2) α has a non-recurrent ray of infinite length.

(3) α has a recurrent ray and a boundary ray.

For, let α be exceptional and not critical. One of the rays at least must have infinite length, as otherwise it would be a cross cut of finite length. Let $|\alpha^+| = \infty$. If α^+ is not recurrent, α is listed under (2); so let α^+ be recurrent. If $|\alpha^-|<\infty$, α is listed under (3), as α^- must be a boundary ray. So let $|\alpha^-|=\infty$. As, by assumption, α is exceptional, α^- cannot be recurrent. Hence α is listed under (2).

We are now going to prove the following

Theorem 13.1. *The exceptional trajectories of a holomorphic quadratic differential φ of finite norm on an arbitrary Riemann surface R cover a set of measure zero. (I.e. of φ-area zero or, equivalently, of Euclidean measure zero in local parameters.)*

Remark. The Riemann sphere and the once and twice punctured Riemann sphere do not carry a non zero holomorphic quadratic differential of finite norm. Except for the torus, the surface must therefore be hyperbolic (i.e. its universal covering surface is the disk); on a torus the statement is of course true.

Since the set of critical trajectories is denumerable, we only have to deal with exceptional trajectories of the second and the third kind. This will be done in two separate sections.

13.2 The φ-area of a general horizontal strip. Let β be an open vertical arc of a meromorphic quadratic differential $\varphi \not\equiv 0$ (i.e. β is not a closed vertical trajectory). The function Φ maps it homeomorphically onto an open vertical interval $\beta' = \Phi(\beta)$ in the w-plane. We continue Φ^{-1} conformally along every horizontal through the points of β', to the left and to the right.

The set of maximal open horizontal intervals through the points of β' along which the continuation is possible is a domain S' of the w-plane, and Φ^{-1} is a (locally injective) conformal mapping of S' into R. For, if $w \in S'$, w is on a horizontal through some point $w_0 \in \beta'$ and the continuation of Φ^{-1} is defined in an open neighborhood of the interval $[w_0, w]$. Also β' is contained in an open set where Φ^{-1} is defined. Therefore S' is open and we can join any two points $w_1, w_2 \in S'$ by a path consisting of horizontal and vertical intervals. The Φ^{-1}-image of S' is the set $S = S(\beta)$ covered by the trajectories through the points of β. Evidently S is also a domain.

Let now E' be a measurable subset of β'. The intersections with S' of the horizontals through the points of E' determine a certain subset of S' which we call the horizontal strip $S'(E')$ defined by E'. Clearly, $S'(E')$ is measurable, and so is its image $S(E) = \Phi^{-1}(S'(E'))$, where $E = \Phi^{-1}(E')$. For, $S'(E')$ can be covered by denumerably many open neighborhoods in which Φ^{-1} is $1-1$-conformal, and therefore $S(E)$ is a denumerable union of measurable sets. Moreover, let $\{U_i'\}$ be an exhaustion of the domain $S'(\beta')$ by denumerably many non overlapping squares U_i' in which Φ^{-1} is injective. We then get, with $B_i' = U_i' \cap S'(E')$, $B_i = \Phi^{-1}(B_i')$

$$\iint_{B_i'} du\,dv = \iint_{B_i} |\Phi'(z)|^2\, dx\,dy = \iint_{B_i} |\varphi(z)|\, dx\,dy.$$

From this we conclude that

$$\begin{aligned}\iint_{S'(E')} du\,dv &= \sum_i \iint_{B_i'} du\,dv = \sum_i \iint_{B_i} |\varphi(z)|\, dx\,dy \\ &\geq \iint_{S(E)} |\varphi(z)|\, dx\,dy \equiv |S(E)|_{\varphi},\end{aligned}$$

since the sets B_i cover $S(E)$. Equality holds iff Φ^{-1} is one-one on $S'(E')$.

13.3 Trajectories with a non recurrent ray of infinite length. Let $\varphi \not\equiv 0$ be a holomorphic quadratic differential of finite norm on an arbitrary Riemann surface R. We consider an oriented closed vertical arc β of φ. As the set of points of R which is covered by closed trajectories of φ is open, the set $E = E(\beta)$ of points $P \in \beta$ which lie on a non closed trajectory is closed. The trajectory through a point $P \in E$ is denoted by $\alpha(P)$, its positive ray (i.e. the ray which leaves β to the right) by $\alpha^+(P)$.

Theorem 13.2. *The set $H \subset E(\beta)$ of points $P \in \beta$ such that the trajectory $\alpha(P)$ has at least one non recurrent ray of infinite φ-length has one dimensional mesure zero.*

Proof. We denote by E_a^+ the set consisting of all points $P \in E(\beta)$ such that $\alpha^+(P)$ carries a closed interval $[P, P']$ of φ-length $a > 0$ which has only P in common with β. Evidently, for $b>a$, $E_b^+ \subset E_a^+$, and the set E^+ of all points $P \in E$ such that $|\alpha^+(P)|_\varphi = \infty$ and $\alpha^+(P) \cap \beta = \{P\}$ is the intersection $E^+ = \bigcap_{n \in \mathbb{N}} E_n^+$.

Let $P \in E_a^+$. Then the interval $[P, P']$ of $\alpha^+(P)$ is the middle diameter of a closed horizontal rectangle S of length a which is schlicht on R and whose intersection with β consists of its initial vertical interval I or of a subinterval of it (Fig. 39). Therefore the rays $\alpha^+(\tilde{P})$ with $\tilde{P} \in I \cap E$ also carry such an interval of length a which has only its initial point $\tilde{P}$ in common with β. We conclude that the set E_a^+ is a relatively open subset of E, for every $a>0$. Therefore E^+ is the intersection of denumerably many measurable subsets of β, hence measurable itself.

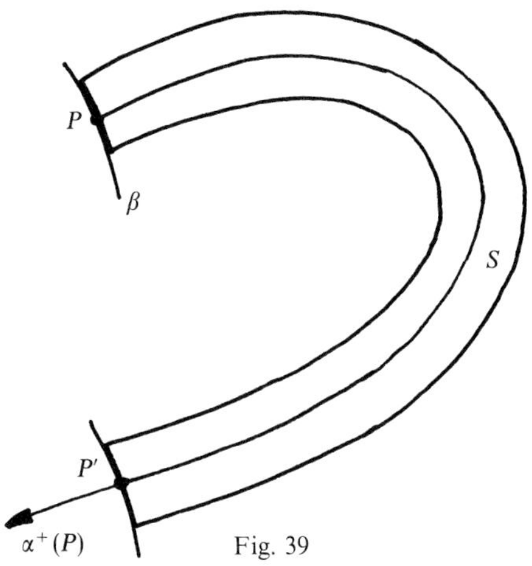

Fig. 39

We now pass to the vertical interval $\beta' = \Phi(\beta)$ and the set $E' = \Phi(E^+)$ on it. The right half strip S' defined by E' and consisting of the right half lines with initial point in E' is of course measurable itself. Moreover it is mapped by Φ^{-1}, or more accurately by the continuation of Φ^{-1} along these half lines, injectively. This follows easily from the fact that the φ-length of the rays $\alpha^+(P)$ is infinite and that they have only their initial point P in common with β, for all $P \in E^+$. Let S denote the set covered by these rays. Then

$$\|\varphi\| \geq |S|_\varphi = |S'|$$

where $|S'|$ is the Euclidean area of S'. As every horizontal in S' has infinite length, we conclude that E' and hence also E^+ is a nullset.

Let now $H^+ = H^+(\beta)$ be the subset of $E(\beta)$ for which $\alpha^+(P)$ has infinite length and is not recurrent. We construct a successive subdivision of β into subintervals of smaller and smaller length, tending to zero. Let $\beta_1, \beta_2, \beta_3, \ldots$ be the denumerably many closed subintervals. Then, every $P \in H^+$ is contained in $E^+(\beta_\nu)$ for some ν. Therefore H^+ is *a* the union of denumerably many nullsets, hence a nullset itself.

Finally, with H^- the set of points $P \in E$ such that $|\alpha^-(P)|_\varphi = \infty$ and $\alpha^-(P)$ non recurrent, $H = H^+ \cup H^-$, which proves the theorem. Clearly, by exhaustion by means of a sequence of closed intervals, the theorem also holds for open vertical arcs β.

13.4 Trajectories with a recurrent ray and a boundary ray

Theorem 13.4. *Let $\varphi \neq 0$ be a holomorphic quadratic differential of finite norm on an arbitrary Riemann surface R and let β be an oriented open vertical arc of φ. Then the set H of points $P \in \beta$ such that $\alpha(P)$ consists of a boundary ray and a recurrent ray is a nullset.*

Proof. We denote by E_0^+ the set of points $P \in E = E(\beta)$ such that $\alpha^+(P)$ is recurrent, while $\alpha^-(P)$ is a boundary ray which has only P in common with β. The first step is to show that E_0^+ is measurable.

Let $P_0 \in E$ and assume that the ray $\alpha^+(P_0)$ has φ-length $> a$ for some positive number a. Then, the subinterval $[P_0, P_1]$ of $\alpha^+(P_0)$ of length a is the central diameter of a closed horizontal rectangle S of length a which is schlicht on R. Let I be the initial vertical side of S. For every $P \in I \cap \beta$ the ray $\alpha^+(P)$ passes through S and hence $|\alpha^+(P)|_\varphi > a$. The set of points $P \in E$ such that $|\alpha^+(P)|_\varphi > n$ is therefore relatively open for every $n \in \mathbb{N}$, and the set of points $P \in E$ with $|\alpha^+(P)|_\varphi = \infty$ is a G_δ-subset of E (denumerable intersection of relatively open sets). Now, every ray $\alpha^+(P)$ which is recurrent has infinite length and we have seen in the preceding section that the set of points $P \in E$ for which $\alpha^+(P)$ has infinite length without being recurrent is a nullset. Therefore the set of $P \in E$ with $\alpha^+(P)$ recurrent is measurable.

By the same argument, the set of points $P \in E$ for which $|\alpha^-(P)|_\varphi \leq a$ is relatively closed, and the set for which $|\alpha^-(P)|$ is finite is therefore an F_σ-subset of E (denumerable union of closed subsets of E). Now, if $\alpha^-(P)$ is a boundary ray, it has either finite length or else it has infinite length without being recurrent. The set of points with the latter property is a nullset, according to the preceding section. We conclude that the set of points $P \in E$ such that $\alpha^-(P)$ is a boundary ray is measurable.

Let $\alpha^-(P_0)$ have at least one other point P_1 in common with β. Then, since β is open, the interval $[P_0, P_1]$ on $\alpha^-(P_0)$ is the middle diameter of a closed horizontal rectangle with both its vertical sides on β. Therefore, for every P which is sufficiently close to P_0 the ray $\alpha^-(P)$ also passes through β after P. We conclude that the corresponding set is open and hence the set of points $P \in E$ with $\alpha^-(P) \cap \beta = \{P\}$ is a relatively closed subset of E.

To sum it up, the set E_0^+ is the intersection of three measurable subsets of E and therefore itself measurable.

We now define a sequence of mappings T_n, $n = 1, 2, 3, \ldots$, of the set E_0^+ onto sets $E_n^+ \subset E$ by the following requirement: For $P_0 \in E_0^+$, $T_n(P_0) = P_n$ is the n-th intersection (after P_0) of $\alpha^+(P_0)$ with β. As β is open and $\alpha^+(P_0)$ is recurrent, there exists a well determined P_n for every $P_0 \in E_0^+$. We define $E_n^+ = T_n(E_0^+)$. The mapping T_n is a bijection: Let $T_n(P_0) = T_n(P_0') = P_n$. The rays $\alpha^+(P_0)$ and $\alpha^+(P_0')$ must cut β at P_n in the same direction, because otherwise $\alpha(P_n)$ would be a cross cut, which is impossible. Therefore one of the rays $\alpha^+(P_0)$ and $\alpha^+(P_0')$ is a subray of the other, and as P_n is the n-th intersection, $P_0 = P_0'$.

It follows in the same way that for $n \neq m$ $E_n^+ \cap E_m^+ = \emptyset$. For, let $P \in E_n^+ \cap E_m^+$. Then, there are points P_0 and P_0' such that $P = P_n = P_m$ is the n-th intersection of $\alpha^+(P_0)$ resp. the m-th intersection of $\alpha^+(P_0')$ with β. Again, the two rays must cut

β at P in the same direction, and one of them, say $\alpha^+(P_0)$, is a subray of the other. If $P_0' \neq P_0$, the ray $\alpha^-(P_0)$ cuts β at P_0', which is impossible, since $\alpha^-(P_0)$ has only P_0 in common with β. Therefore $P_0' = P_0$. The point P is the n-th intersection and at the same time the m-th intersection of $\alpha^+(P_0)$ with β, which is only possible if $m = n$, contradicting the assumption $m \neq n$.

We now show that E_n^+ is measurable and $|E_n^+|_\varphi = |E_0^+|_\varphi$. On every ray $\alpha^+(P_0)$, $P_0 \in E_0^+$, we consider the interval $[P_0, P_n]$, $P_n \in E_n^+$. We leave those initial points P_0 away (there are at most two of them) for which the subinterval $[P_0, P_n]$ of $\alpha^+(P_0)$ passes through one of the endpoints of β. For every other P_0 there exists a schlicht open horizontal rectangle S of φ-length $|[P_0, P_n]|_\varphi$ which contains the open interval (P_0, P_n) as a diameter and which is traversed by β exactly $n-1$ times (Fig. 40).

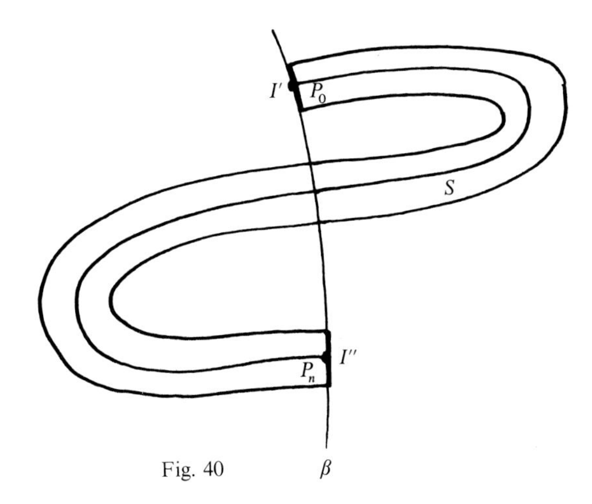

Fig. 40

The set of basic vertical intervals I' of these horizontal strips S cover E_0^+ up to the (at most two) exceptional points. For every I' the subset $I' \cap E_0^+$ is mapped by T_n onto the corresponding set $I'' \cap E_n^+$ in a measure preserving way, where I'' is the top vertical interval of S. Let J be one of the components of the open covering $\bigcup I'$ of E_0^+. There are denumerably many intervals I'_ν which cover J (every closed subinterval of J is covered by finitely many, and J can be exhausted by denumerably many closed subintervals). We can therefore partition $J \cap E_0^+$ into denumerably many subsets each of which is mapped by T_n in a measure preserving way and the images of the different subsets are disjoint. Therefore T_n is a measure preserving map of $J \cap E_0^+$ onto $T_n(J \cap E_0^+)$. Adding up all the denumerably many components of the open covering we find that E_n^+ is a denumerable union of measurable sets, hence measurable, and that $|E_0^+|_\varphi = |E_n^+|_\varphi$.

We conclude that $|E_0^+|_\varphi = |E_1^+|_\varphi = |E_2^+|_\varphi = \dots$, and as all these sets are disjoint subsets of the vertical interval β of finite φ-length, their φ-length is zero. Similarly, $|E_0^-|_\varphi = |E_1^-|_\varphi = |E_2^-|_\varphi = \dots = 0$. This means that all these sets are nullsets, and so is their union $\bigcup_0^\infty (E_i^+ \cup E_i^-)$.

Let now $P \in \beta$ be a point such that the trajectory $\alpha(P)$ is composed of a boundary ray and a recurrent ray. Then, the boundary ray has a last intersection P_0 with β, which is either a point of E_0^+ or of E_0^-. Therefore P is in some set E_i^+ or E_i^-, $i=0,1,2,\ldots$. The set H of all such points P is thus equal to the union $H=\bigcup_0^\infty (E_i^+ \cup E_i^-)$, hence a nullset.

The proof of Theorem 13.1 is now an easy consequence of Theorems 13.3 and 13.4. Let (P_n) be a sequence of regular points of φ which is everywhere dense on R. Let β_n be the largest open vertical interval with midpoint P_n and of length ≤ 1, say. The set of exceptional trajectories which have a point in common with β_n has φ-area zero, for every n. On the other hand, every exceptional trajectory of φ cuts some β_n and therefore lies in the union of denumerably many nullsets.

13.5 Strips of horizontal cross cuts of finite length. In order to apply the length area method to quadratic differentials of finite norm on arbitrary Riemann surfaces it is necessary to partition the surface into certain subsets of trajectories. These are, besides the ring domains swept out by closed trajectories, strips of horizontal cross cuts and sets of spirals (as nearly closed trajectories).

Definition 13.5. Let φ be a holomorphic quadratic differential of finite norm on an arbitrary Riemann surface R. Let β be a vertical interval with respect to φ on R. The set of horizontal cross cuts of finite length which cut β exactly once is called the horizontal strip $S=S(\beta)$ determined by β.

We denote by $E=E(\beta)$ the set of points P of S on β. The trajectory $\alpha(P)$ is characterized by the fact that both rays $\alpha^+(P)$ and $\alpha^-(P)$ are boundary rays of finite length which have only P in common with β. According to the preceding section the two sets of points P with the above properties are both measurable. Therefore the same is true for their intersection which is the set E.

The set $E'=\Phi(E)$ on the vertical interval $\beta'=\Phi(\beta)$ is then also measurable. It determines a measurable substrip S' of the general horizontal strip associated with β (Section 13.2). Φ^{-1} is a $1-1$-conformal mapping of S' onto S. For, let the two points $P_1=\Phi^{-1}(w_1)$ and $P_2=\Phi^{-1}(w_2)$ coincide (Fig. 41). Then, the two

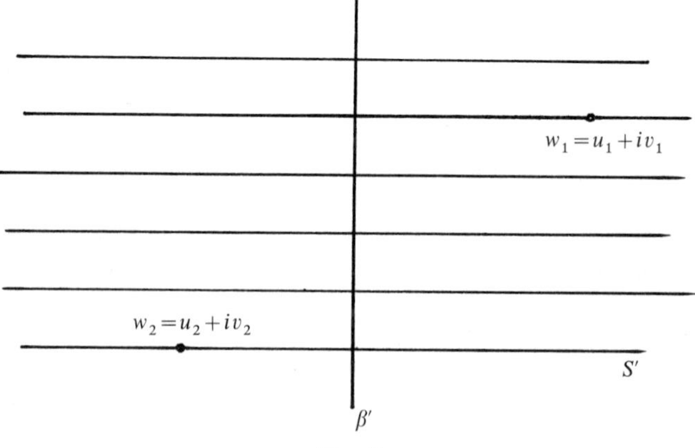

Fig. 41

horizontals through w_1 and w_2 are mapped by Φ^{-1}, onto the same trajectory α through $P_1=P_2$. Since α has only one point P in common with β and Φ is a homeomorphism of β onto β', the two lines through w_1 and w_2 coincide: $v_1 = v_2$. But again, since Φ is a homeomorphism on α (α is not closed), $u_1=u_2$, hence $w_1=w_2$. We conclude that the Euclidean measure of S' is equal to the φ-area of S (Section 13.2).

The set of all horizontal cross cuts of finite φ-length of R can now be partitioned into disjoint horizontal strips. The cross cuts of infinite length can be neglected, as they cover a set of zero area.

To this end, pick a denumerable, everywhere dense set of regular points $P_1, P_2, P_3, \ldots$ on R. To each P_n we assign the sequence of open vertical intervals β_{nk} of φ-length $\frac{1}{k}$, $k \geq k_0(P_n)$, with midpoint P_n. The number $k_0(P_n) \in \mathbb{N}$ is chosen smallest possible, i.e. we start with the largest possible vertical interval of φ-length ≤ 1 for every P_n.

The horizontal strips S_{nk} associated with the intervals β_{nk} contain all the horizontal cross cuts of finite φ-length. For, let α be such a trajectory and let $P \in \alpha$. There exists a horizontal square $Q(P)$ with center P through which α passes only once. We can evidently find an interval β_{nk} in $Q(P)$ which cuts α. Therefore α belongs to the strip S_{nk} associated with β_{nk}.

In order to get a sequence of disjoint horizontal strips, let $\beta_1, \beta_2, \beta_3, \ldots$ be a renumbering of the vertical intervals. Let S_1 and S_2 be the horizontal strips associated with β_1 and β_2 respectively. We set $E_2=\beta_2 \cap S_2$, and we want to show that the set of points $P_2 \in E_2$ for which $\alpha(P_2)$ belongs to S_1 is measurable. If $\alpha(P_2)$ cuts β_1 at least once, the same is true for every $P \in E_2$ in a neighborhood of P_2. Therefore this set is relatively open. The same holds for the points $P_2 \in E_2$, for which $\alpha(P_2)$ cuts β_1 at least twice. Therefore the set of points $P \in E_2$ for which $\alpha(P)$ belongs to S_1, i.e. cuts β_1 exactly once, is the difference of two relatively open subsets of E_2 hence measurable. The intersection $S_1 \cap S_2$ is therefore a measurable substrip of S_2, and so is the difference $S_2 - S_1$. The argument can be repeated. The sequence of strips

$$S_1, S_2 \setminus S_1, S_3 \setminus S_1 \setminus S_2, \ldots$$

with the general term $S_n \setminus S_1 \setminus S_2 \setminus \ldots \setminus S_{n-1}$ provides the desired exhaustion.

13.6 Spiral sets

Definition 13.6. For a given holomorphic quadratic differential φ of finite norm on an arbitrary Riemann surface R, let β be an oriented vertical interval. By the spiral set determined by β we mean the set of spirals α (Section 13.1c) which have a non empty intersection with β. We denote it by $\mathscr{S}(\beta)$. The same notation is used for the union of these spiral trajectories as a point set of R.

The set $E=\beta \cap \mathscr{S}(\beta)$ is measurable. For, the set of points $P \in \beta$ such that $\alpha(P)$ is not closed and $\alpha^+(P)$ has infinite length is a G_δ-subset of a relatively closed set. Now, every ray $\alpha^+(P)$ of infinite length is either recurrent or not. The set of points $P \in \beta$ where the second case occurs has measure zero (Section 13.3), therefore the set E^+ of points $P \in \beta$ with recurrent $\alpha^+(P)$ is measurable.

The same is true for the set E^- of points $P \in \beta$ with a recurrent ray $\alpha^-(P)$. As $E = E^+ \cap E^-$, it is measurable.

We are now going to partition the set $\mathscr{S}(\beta)$ into measurable strips of the first kind and pairs of measurable strips of equal width of the second kind. The first step consists in a covering of $\mathscr{S}(\beta)$ by rectangular strips based on β.

Every spiral $\alpha \subset \mathscr{S}(\beta)$ is subdivided by its intersection points with β into denumerably many subintervals α_n. We say that α_n is of the first kind, if its two endpoints lie on different edges of β; otherwise, α_n is said to be of the second kind. In the last case we denote it by α_n^+ if both its endpoints lie on the edge β^+ of β, otherwise by α_n^-.

To fix the ideas, let β be open and disregard the at most two intervals passing through the endpoints of β. Then, every interval α_n is contained in a rectangular strip S, i.e. the Φ^{-1}-image of a horizontal rectangle S': $u_0 \leq u \leq u_0 + a_n$, $v_0 < v < v_1$ of the $w = u + iv$-plane, with $a_n = |\alpha_n|_\varphi$ the φ-length of the interval α_n. Its vertical sides are subintervals of β, but they lie on the two different edges β^+ and β^- of β. Otherwise S has no point in common with β. There clearly exists a lagest such rectangular strip (the union of all of them): It is called the rectangular strip S_n associated with the interval α_n. The same is true for the intervals α_n^+ and α_n^- of the second kind: They determine strips S_n^+ and S_n^- respectively which are schlicht on R and which have both their vertical sides on β^+ or on β^-.

The above construction is performed for every spiral in $\mathscr{S}(\beta)$. Let α_n and $\tilde{\alpha}_m$ be two intervals on spirals α and $\tilde{\alpha}$ in $\mathscr{S}(\beta)$ respectively. Assume that their rectangular strips S_n and $\tilde{S}_m$ have an interior point P in common. Then the two strips have a common horizontal interval through P, and since they are both maximal, they coincide. We thus get a well determined, denumerable set of non overlapping rectangular strips of the first and of the second kind which cover $\mathscr{S}(\beta)$.

In the next step we pass to the intersections $S_n \cap \mathscr{S}(\beta)$ and combine the strips of the second kind to pairs of equal width. This will then be the final strip decomposition of $\mathscr{S}(\beta)$. To this end, consider an arbitrary rectangular strip S_n of the above collection. The intersection $S_n \cap \mathscr{S}(\beta)$ consists of all the subintervals of the spirals of $\mathscr{S}(\beta)$ which are in S_n. This is the Φ^{-1}-image of a certain set of horizontal straight intervals in the Euclidean rectangle S_n'. Let the measure of the ordinates of these horizontals be denoted by b_n. Then $a_n b_n$ is the φ-measure of the set $S_n \cap \mathscr{S}(\beta)$ of spiral intervals of $\mathscr{S}(\beta)$ in S_n. The total φ-measure of $\mathscr{S}(\beta)$ is the sum

$$\|\varphi\|_{\mathscr{S}(\beta)} = \sum a_n b_n + \sum a_n^+ b_n^+ + \sum a_n^- b_n^-,$$

i.e. the sum of the areas of all intersections of $\mathscr{S}(\beta)$ with the φ-rectangles S_n, S_n^+, S_n^-.

The set E has two identical replicas E^+ and E^- on the two edges β^+ and β^- respectively. Every strip S_n has one of its vertical sides on β^+, the other on β^-, while every $S_n^+ (S_n^-)$ has both sides on $\beta^+ (\beta^-)$. We therefore get for the φ-measures

$$|E| = |E^+| = \sum b_n + 2\sum b_n^+ = |E^-| = \sum b_n + 2\sum b_n^-,$$

hence $\sum b_n^+ = \sum b_n^-$. In other words, the total width of the strips $S_n^+ \cap \mathscr{S}(\beta)$ is the same as that of the strips $S_n^- \cap \mathscr{S}(\beta)$. This fact allows, after a horizontal subdivision of the strips of the second kind, a pairing $S_n^+ \leftrightarrow S_n^-$ such that $b_n^+ = b_n^-$. One simply starts with the strips with largest ordinate sets, say (after renumbering) $S_1^+ : b_1^+$ and $S_1^- : b_1^-$. If $b_1^+ > b_1^-$, we cut off a part $\tilde{S}_1^+$ of S_1^+, by a horizontal arc in S_1^+, such that $\tilde{b}_1^+ = b_1^-$. The first pair now is $\tilde{S}_1^+$, S_1^- respectively their intersections with $\mathscr{S}(\beta)$. We continue likewise with the rest. The process has at most denumerably many steps.

What is the intersection of two spiral sets determined by vertical intervals β_1 and β_2? Let E_1, E_2 be the sets of points on spiral trajectories on β_1 and β_2 respectively. We want to show that the set E_2' of points $P_2 \in E_2$ for which $\alpha(P_2) \subset \mathscr{S}_1 \cap \mathscr{S}_2$ is measurable. But $\alpha(P_2)$ is a spiral in $\mathscr{S}_2$ which cuts β_1. Then, the same is true for all $P \in E_2$ in a certain neighborhood of P_2. Therefore E_2' is a relatively open subset of E_2, hence measurable. So is $E_2 \setminus E_2'$. The spirals through this set are the spirals of $\mathscr{S}_2 \setminus \mathscr{S}_1$. Instead of working with all the spirals through β_2 (i.e. through E_2), we can work with the set $\mathscr{S}_2 \setminus \mathscr{S}_1$. The procedure of partitioning into strips is the same.

Chapter V. The Metric Associated with a Quadratic Differential

Introduction. The invariant line element $|\varphi(z)|^{1/2}|dz|$ was introduced in Section 5.3 and the local properties of the corresponding metric were investigated in Sections 5.4 and 8. In this chapter, we study its global properties.

Given a curve $\gamma\colon t \to P(t)$, where the parameter t varies in some interval I, one can compare the φ-length of small subarcs $\gamma([t_1, t_2])$ with the φ-length of other subarcs joining its two endpoints $P_i = \gamma(t_i)$ $i = 1, 2$. The varied arcs can be restricted to be near $\gamma([t_1, t_2])$, or homotopic to it. For a holomorphic φ we have shown a general existence and uniqueness theorem (Sections 5.4 and 8.1): any two points P_1 and P_2 which are sufficiently close to each other can be joined by a unique shortest arc (without any restrictions for the variations).

A geodesic is defined to be locally shortest (Definition 5.5.1). Section 14.1 deals with the uniqueness of geodesics. The proofs are based on Teichmüller's lemma (Theorem 14.1), which uses the argument principle and the rotation of a tangent vector along a simple closed loop. The theorem has a striking application to domains of connectivity ≤ 3 (there are no recurrent trajectory rays). The minimal length (in the large) of a geodesic arc is shown by a geometric proof in § 16. This yields another uniqueness proof for geodesic connections. In § 17, the analogous reasonings are carried out for closed geodesics. The chapter is completed by § 18 which deals with the existence of geodesics.

§ 14. Uniqueness of Geodesics

14.1 Teichmüller's lemma (Teichmüller [3], p. 162; Ahlfors [2]). We first state the argument principle in its general form. Let D be a relatively compact domain of the z-plane bounded by finitely many piecewise smooth curves and let φ be meromorphic in $\bar{D}$. The zeroes and poles of φ in D are denoted generically by z_i, with n_i and $-n_i$, respectively, their orders. Likewise z_j denotes a zero (pole) of φ on ∂D, with order n_j $(-n_j)$. The points z_j subdivide the boundary of D into arcs Γ_j. The interior angle between the two adjacent arcs at the point z_j is denoted by ϑ_j $(0 \leq \vartheta_j \leq 2\pi)$. Then

$$(1) \qquad \frac{1}{2\pi}\int_\Gamma d\arg\varphi(z) := \frac{1}{2\pi}\sum_j \int_{\Gamma_j} d\arg\varphi(z) = \sum_i n_i + \sum_j \frac{\vartheta_j}{2\pi} n_j.$$

Note that the integral is defined to be taken over the individual open arcs Γ_j, where $\arg\varphi(z)$ is defined. The critical points on the boundary contribute with the weight of the interior angle compared to the full angle.

Definition 14.1. A geodesic polygon with respect to a (meromorphic) quadratic differential φ or simply a φ-polygon is a curve Γ composed of open straight arcs (Definition 5.5.2) and their endpoints (the endpoints can be critical points of φ). It is called simple, if it is a Jordan arc, and simple closed, if it is a Jordan curve.

Theorem 14.1 (Teichmüller's lemma). *Let D be the interior of a simple closed φ-polygon Γ with sides Γ_j (=open straight arcs) and interior angles ϑ_j $(0 \leq \vartheta_j \leq 2\pi)$ at its vertices. The quadratic differential φ is supposed to be meromorphic in $\bar{D}$ (the only points on Γ which can be critical are the vertices). Then, with the previous notation, we have*

$$\sum_j \left(1 - \vartheta_j \frac{n_j+2}{2\pi}\right) = 2 + \sum_i n_i. \tag{2}$$

Proof. Along the sides Γ_j of Γ we have $\arg \varphi(z)\, dz^2 = \text{const.}$ (the constant depending on j), hence

$$d(\arg \varphi(z)\, dz^2) = d \arg \varphi(z) + 2\, d(\arg dz) = 0.$$

The argument of the tangential vector dz increases by $2\pi - \sum_j (\pi - \vartheta_j)$ along the union of the sides Γ_j. We thus get

$$\frac{1}{2\pi} \int_\Gamma d \arg \varphi(z) = -\frac{2}{2\pi} \int_\Gamma d(\arg dz), \tag{3}$$

and therefore

$$\sum_i n_i + \sum_j \frac{\vartheta_j}{2\pi} n_j = -2 + \sum_j \left(1 - \frac{\vartheta_j}{\pi}\right).$$

At each corner we have $1 - \frac{\vartheta_j}{\pi} - \frac{\vartheta_j}{2\pi} n_j = 1 - \vartheta_j \frac{n_j+2}{2\pi}$, which proves the theorem. If, in particular, φ is holomorphic in D, all the n_i are non negative, hence $\sum_i n_i \geq 0$, which involves

$$\sum_j \left(1 - \vartheta_j \frac{n_j+2}{2\pi}\right) = 2 + \sum n_i \geq 2. \tag{4}$$

Remark. Teichmüller's lemma has a version for doubly connected domains. Suppose G is a doubly connected domain of the z-plane bounded by the two simple φ-polygons Γ' and Γ'', positively oriented with respect to G. The argument principle (1) is the same, with $\Gamma = \Gamma' + \Gamma''$ and ϑ_j the interior angles at the vertices of Γ, $0 \leq \vartheta_j \leq 2\pi$. The rotation of the tangential vector along the sides of the inner boundary curve is

$$-[2\pi - \sum_j \{\pi - (2\pi - \vartheta_j)\}] = -2\pi - \sum_j (\pi - \vartheta_j),$$

the term in the square bracket being the rotation of dz in a counter clockwise circuit of the curve. We therefore find, using the relation (3),

(5)
$$\sum_i n_i + \sum_j \frac{\vartheta_j}{2\pi} n_j = -\frac{1}{\pi} \int_\Gamma d(\arg dz)$$
$$= \frac{1}{\pi} \sum_j (\pi - \vartheta_j) = \sum_j \left(1 - \frac{\vartheta_j}{\pi}\right),$$

and hence

(6)
$$\sum_j \left(1 - \vartheta_j \frac{n_j + 2}{2\pi}\right) = \sum_i n_i \geq 0,$$

with n_i the orders of the interior zeroes of φ in G.

14.2 Uniqueness of geodesic arcs. The above relation, combined with the angle condition $\vartheta_j \geq \frac{2\pi}{n_j + 2}$ (Theorem 8.1) allows to prove the uniqueness of geodesic connections between two points.

Theorem 14.2.1. *Let φ be a holomorphic quadratic differential in a simply connected domain G. Then any two points of G can be joined by at most one geodesic arc.*

Proof. Let γ_1 and γ_2 be two geodesics joining z_1 and z_2 in G. If they do not coincide, we can find two subarcs joining a point z' to a point z'' and forming, together, a Jordan curve γ (Fig. 42). The angle condition is satisfied except

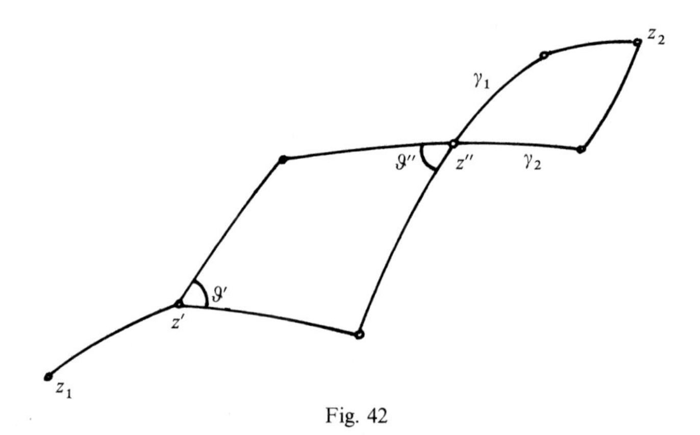

Fig. 42

possibly for the two intersections z' and z'', where the interior angles ϑ' and ϑ'' of γ satisfy $0 < \vartheta', \vartheta'' < 2\pi$. So, in the sum (2) we can get at most two positive terms and both are smaller than one, contradicting (2).

The theorem implies, as special cases, that a holomorphic quadratic differential φ in a simply connected region does not allow a closed geodesic nor a geodesic loop ($z_1 = z_2$). This evidently is still true if we allow φ to have at most one simple pole inside of the closed loop.

The theorem can be generalized to homotopic geodesic arcs on arbitrary Riemann surfaces.

Theorem 14.2.2. *Let φ be a holomorphic quadratic differential on an arbitrary Riemann surface R. Then, in every homotopy class of curves joining two points P_1 and P_2 of R there is at most one geodesic. In particular, for $P_1 = P_2 = P$ there is no closed geodesic or geodesic loop which is contractible to P.*

Proof. Let γ_1 and γ_2 be homotopic geodesic arcs joining P_1 and P_2. Let $\hat{R}$ be the universal covering surface of R and choose a lift z_1 of P_1 on $\hat{R}$. By the monodromy theorem, the lifts $\hat{\gamma}_1$ and $\hat{\gamma}_2$ of γ_1 and γ_2 respectively with initial point z_1 have the same terminal point z_2 above P_2. Moreover, they are geodesic arcs of the lift $\hat{\varphi}$ of φ. Therefore $\hat{\gamma}_1 = \hat{\gamma}_2$ and hence $\gamma_1 = \gamma_2$.

14.3 Uniqueness of closed geodesics

Theorem 14.3. *Suppose γ is a closed geodesic loop of a holomorphic quadratic differential φ on an arbitrary Riemann surface R. Then, γ is uniquely determined in its free homotopy class except for the case where it is a power of a closed θ-trajectory or of a boundary curve of an annulus swept out by closed θ-trajectories.*

Proof. Let $\hat{R}(\gamma)$ be the annular covering surface of R associated with γ. We denote by $\hat{\gamma}$ a closed lift of γ to $\hat{R}(\gamma)$, oriented in such a way that it has winding number one around the inner boundary component of $\hat{R}(\gamma)$. Since it is a geodesic of the lift $\hat{\varphi}$ of φ, it is a simple closed loop.

Suppose γ_1 is a geodesic which is freely homotopic to γ on R. Let $\hat{\gamma}_1$ be a closed lift of γ_1. If $\hat{\gamma}_1 \cap \hat{\gamma} \neq \emptyset$, $\hat{\gamma}_1 = \hat{\gamma}$, and hence $\gamma_1 = \gamma$ by projection. The equality follows from Theorem 14.2.1 by passing to the universal covering surface of the annulus.

Let now $\hat{\gamma}_1 \cap \hat{\gamma} = \emptyset$. Then, the two curves in $\hat{R}(\gamma)$ bound a doubly connected domain $\hat{G} \subset \hat{R}(\gamma)$. Applying Teichmüller's lemma for doubly connected domains (see remark at the end of Section 14.1) we find that $\hat{\varphi}$ cannot have any zeroes in $\hat{G}$ and that all the interior angles ϑ_j at the vertices z_j of the boundary curves must be minimal, i.e. $\vartheta_j = \dfrac{2\pi}{n_j + 2}$, where n_j is the order of the zero z_j of φ. In particular, the directions of all the straight segments of $\hat{\gamma}$, say, are the same: $\arg \hat{\varphi}(z)\,dz^2 = \theta$. In the interior neighborhood of $\hat{\gamma}$ there are closed θ-trajectories. No such θ-trajectory can have a point in common with $\hat{\gamma}_1$, unless it coincides with it (by the first part of the proof). We conclude that $\hat{\gamma}$ and $\hat{\gamma}_1$ are parallel closed θ-trajectories or boundary curves of the subannulus $\hat{R}_0$ of $\hat{R}(\gamma)$ swept out by these θ-trajectories.

Assume $\hat{\gamma}_2$ is a closed θ-trajectory of $\hat{\varphi}$ in $\hat{R}_0$. Its projection γ_2 is a closed θ-trajectory of φ, and hence a Jordan curve or a certain power of it. The projection R_0 of $\hat{R}_0$ into R is an annulus swept out by θ-trajectories. The geodesic γ is one of these or a boundary curve of R_0, possibly a power of it (i.e. the same curve, run through several times).

We find in particular: if γ is not freely homotopic to a power of a simple closed loop and γ is a geodesic, it is uniquely determined in its free homotopy class.

§ 15. Domains of Connectivity ≤ 3

15.1 Geodesic arcs in simply connected domains. In Section 14.2 we have seen that for a holomorphic quadratic differential in a simply connected domain there are no closed geodesics. In this § Teichmüller's lemma is used to give more precise information about the geodesics in domains up to connectivity three.

Theorem 15.1. *Every maximal geodesic arc (and in particular every non critical trajectory) of a holomorphic quadratic differential in a simply connected region is a cross cut* (Section 10.2).

Proof. Let φ be holomorphic in the simply connected region G and let γ be a maximal geodesic ray of φ with initial point z_0 (γ is not properly contained in another such ray). We use the φ-length u of γ as parameter, $0 \leq u < u_\infty$. Assume that $\gamma(u)$ does not tend to the boundary Γ of G for $u \to u_\infty$. Then there is a sequence $u_n \to u_\infty$ such that $z_n = \gamma(u_n) \to z$ for some $z \in G$. The point z has a relatively compact open neighborhood U such that every geodesic arc in U can be continued to the boundary of U. Let γ_n be the maximal subarc of γ which contains z_n and lies in U. There is a point $z_m \notin \gamma_n$, and therefore the subarc $[z_n, z_m]$ on γ is not contained in U. On the other hand, the points z_n and z_m can be joined by a geodesic arc $\tilde{\gamma} \subset U$ if they are sufficiently near z. This contradicts the uniqueness of geodesic connections in G. We have thus seen that every subray of a maximal geodesic arc tends to the boundary.

15.2 Exclusion of recurrent trajectory rays. In a simply connected domain every geodesic ray is a boundary ray; in particular, there are no recurrent trajectory rays (see Section 10.2). The latter fact is true for all domains of connectivity ≤ 3.

Theorem 15.2 (Jenkins [1], [5], Kaplan [1]). *No trajectory ray of a holomorphic quadratic differential φ in a domain G of connectivity ≤ 3 is recurrent.*

Proof. The proof follows an idea of W. Kaplan. Let α^+ with initial point z_0 be a recurrent trajectory ray. As we can move z_0 along α^+, we can assume that z_0 is a regular point of φ. Let β be an interval of the vertical trajectory through z_0, with one of its endpoints z_0, which is cut infinitely often by α^+. If the next intersection z_1 of α^+ with β after z_0 is in the positive direction (Fig. 43), we look at the Jordan curve γ which is composed of the two subarcs α_0 and β_0 of α^+ and β respectively between the points z_0 and z_1. In the opposite case, we wait for the next intersection of α^+ with β_0 after z_1 (Fig. 44). It must be in the positive direction, as γ is a Jordan curve. We therefore only have to consider the first case. We continue α^+ through z_1 and call the subdomain of $\hat{\mathbb{C}} \setminus \gamma$ which it enters at z_1 the interior G_0 of γ. Let z_2 be the next intersection of α^+ with the subinterval β_0 (Fig. 45). The arc γ_1 on α^+ between z_1 and z_2 is a cross section of G_0 which splits it into two subdomains

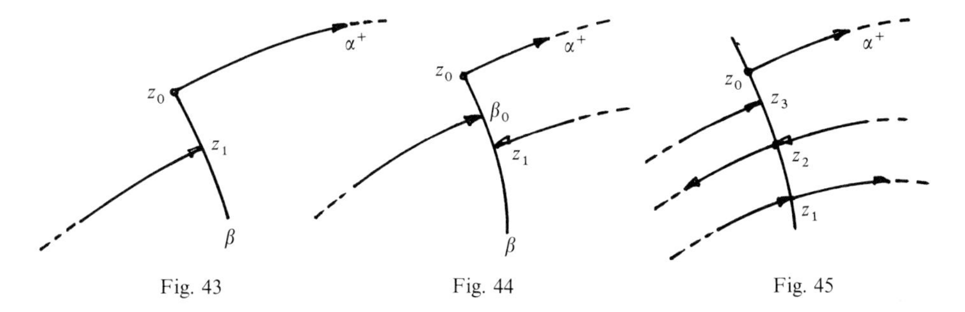

Fig. 43 Fig. 44 Fig. 45

G_1 and G_2, with $z_0 \in \partial G_1$. We can apply Teichmüller's lemma (Theorem 14.1) to these domains: it does not matter whether they contain the point at infinity or not nor if the boundary passes through ∞, as is seen by a linear fractional transformation. The interior angles at the vertices of G_1 are $\frac{\pi}{2}$ at z_0 and z_2 and π at z_1; the ones of G_2 are both $\frac{\pi}{2}$. The left hand side of (2) becomes one (all the n_j are equal to zero) for both domains. If they would not contain a boundary component of G, we would get the contradiction that the right hand side of (2) is ≥ 2.

Now we continue α^+ beyond z_2 to its next intersection z_3 with β_0. The subarc δ_1 of α^+ between z_2 and z_3 partitions the exterior D_0 of γ into domains D_1 and D_2, with $z_0 \in \partial D_1$. The interior angles at the vertices of D_1 are $\frac{3\pi}{2}$ at z_0 and $\frac{\pi}{2}$ at z_1, z_2 and z_3, and $\frac{\pi}{2}$ at both vertices z_2 and z_3 of D_2. We arrive at the same contradiction unless D_1 and D_2 both contain boundary components of G. But as G has at most three boundary components, this is impossible, which proves the theorem.

§ 16. Minimum Length Property of Geodesic Arcs

16.1 Minimal length in simply connected domains. A geodesic is defined to be locally shortest (Definition 5.5.1). We can now show that it is also the unique globally shortest connection between its endpoints.

Theorem 16.1. *Let $\varphi \not\equiv 0$ be a holomorphic quadratic differential in a simply connected domain G and let γ be a geodesic arc connecting z_0 and z in G. Then the φ-length $|\tilde{\gamma}|_\varphi$ of any arc $\tilde{\gamma} \neq \gamma$ which connects z_0 and z_1 in G is larger than $|\gamma|_\varphi$.*

Proof. Picking finitely many points on $\tilde{\gamma}$ and replacing the arcs between neighboring points by shortest arcs we can assume, without restricting the generality, that $\tilde{\gamma}$ is a geodesic polygon. We choose a Jordan domain G_0 which is bounded by a geodesic polygon Γ_0 and which contains γ and $\tilde{\gamma}$.

First, let γ be a straight arc, possibly with critical endpoints. We can assume that it is a horizontal arc. We now mark those points z_i on γ which lie on a critical vertical arc with respect to G_0. As there are only finitely many zeroes of φ in G_0 and as any vertical arc has at most one intersection with γ, there are only finitely many marked points (including the endpoints) $z_0, z_1, z_2, \ldots, z_n = z$. Let z' be a point of the open subinterval (z_{i-1}, z_i) of γ.

The vertical arc β through z' is a cross cut of G_0 and therefore subdivides it into two simply connected domains, one of which contains z_0, the other contains z (Fig. 46). Therefore $\tilde{\gamma}$ cuts β. The set of all points in G_0 on the

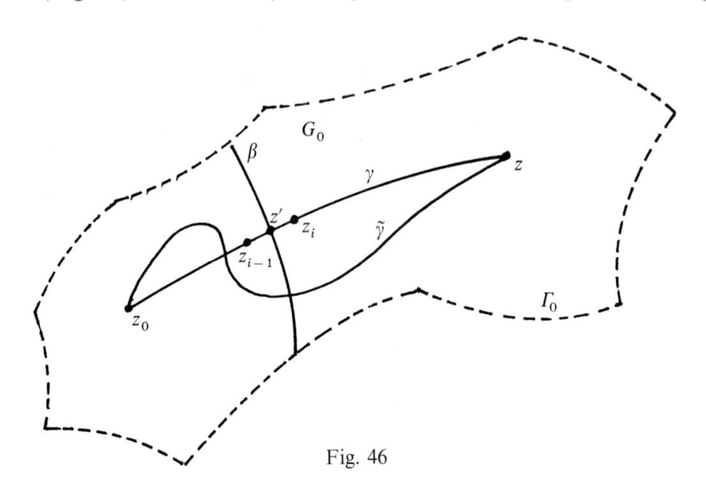

Fig. 46

(connected) vertical arcs through the points $z' \in (z_{i-1}, z_i)$ is a simply connected domain S_i. It is mapped by Φ onto a vertical parallel strip S_i' of width a_i, say. The intersection $\tilde{\gamma} \cap S_i$ consists of finitely many analytic arcs, and their total φ-length clearly is at least equal to a_i, with equality if and only if it is a single arc parallel to γ. As the S_i do not overlap, we conclude that $|\tilde{\gamma}|_\varphi \geq \sum a_i = |\gamma|_\varphi$. Let $|\tilde{\gamma}|_\varphi = |\gamma|_\varphi$. Then $\tilde{\gamma}$ can traverse every vertical strip S_i only once and it must do it on a horizontal arc. Therefore, beginning with $i=1$, $\tilde{\gamma}$ must coincide with γ on the interval $[z_0, z_1]$, then on $[z_1, z_2]$ and so forth: we have $\tilde{\gamma} = \gamma$.

We now pass to an arbitrary geodesic arc γ. It is composed of straight arcs γ_i. For every γ_i we construct the orthogonal strips S_{ij} as before. Any arc β_i orthogonal to γ_i does not meet γ again, because of the uniqueness of geodesic connections. Moreover a β_i which is orthogonal to γ_i cannot intersect a β_k orthogonal to γ_k: The result would be a simple closed geodesic polygon with two interior angles of $\frac{\pi}{2}$, another positive angle at the intersection of β_i with β_k, and the rest satisfying the angle condition, contradicting Teichmüller's lemma. We conclude that $\tilde{\gamma}$ traverses every S_{ij}, which gives the length inequality. Equality can only hold if the intersection of $\tilde{\gamma}$ with S_{ij} consists of a single interval parallel to γ_i, for every i and j. We conclude, as before, that $\tilde{\gamma} = \gamma$.

16.2 Minimum length property on Riemann surfaces. A geodesic arc γ is also the unique shortest connection between its endpoints on a Riemann surface, if compared to curves in its homotopy class only. The same is true if φ is meromorphic and γ does not go through any poles of φ, if we puncture R at the poles and consider the homotopy on the punctured surface $\dot{R}$.

Theorem 16.2. *Let φ be a meromorphic quadratic differential on an arbitrary Riemann surface R and let γ be a geodesic arc joining a point P_0 to a point P and not going through a pole of φ (of course it does not go through a pole of order ≥ 2). Let $\dot{R}=R \setminus \{$poles of $\varphi\}$. Then, for every γ_1 joining P_0 to P which is homotopic to γ on $\dot{R}$ we have $|\gamma_1|_\varphi \geq |\gamma_\varphi$, with equality iff $\gamma_1=\gamma$.*

Proof. Let $\hat{R}$ be the universal covering surface of R, with $\hat{\varphi}$ the lift of φ. The two lifts $\hat{\gamma}$ and $\hat{\gamma}_1$ of γ and γ_1 with the same initial point $\hat{P}_0$ above P_0 have the same endpoint $\hat{P}$ above P. We have $|\gamma_1|_\varphi=|\hat{\gamma}_1|_{\hat{\varphi}} \geq |\hat{\gamma}|_{\hat{\varphi}}=|\gamma|_\varphi$, by the previous theorem, with equality iff $\gamma_1=\gamma$.

16.3 The divergence principle (Teichmüller [2], Strebel [10], [14]). The minimum length property is true under more general circumstances than it was stated in Theorem 16.1. The simplest situation is the following: If γ is a horizontal arc and β_0 and β are the vertical trajectories through its endpoints z_0 and z respectively, then every arc $\tilde{\gamma}$ which joins β_0 and β has φ-length $\geq |\gamma|_\varphi$, with equality if and only if the four arcs bound a rectangle with respect to φ (Fig. 47). One can say that the vertical trajectories through the endpoints of a horizontal arc diverge towards the boundary. We state the fact in the following general form.

Theorem 16.3 (Divergence Principle). *Let $\varphi \not\equiv 0$ be a holomorphic quadratic differential in a simply connected region G and let γ be a geodesic arc of φ, with endpoints z_0 and z. If they are zeroes of φ, their multiplicity is denoted by n_0 and n respectively. Let δ_0 and δ be geodesic arcs through the endpoints of γ, forming angles $\vartheta_0 \geq \pi/(n_0+2)$ and $\vartheta \geq \pi/(n+2)$ with γ respectively. Then the φ-length $|\tilde{\gamma}|_\varphi$ of any arc $\tilde{\gamma}$ joining δ_0 to δ is at least $|\gamma|_\varphi$. Equality holds if and only if $\tilde{\gamma}$ is a geodesic arc forming angles $\tilde{\vartheta}=\pi/(\tilde{n}_0+2)$ and $\tilde{\vartheta}=\pi/(\tilde{n}+2)$ with the arcs δ_0 and δ respectively* (Fig. 48) *and such that all the interior angles of the*

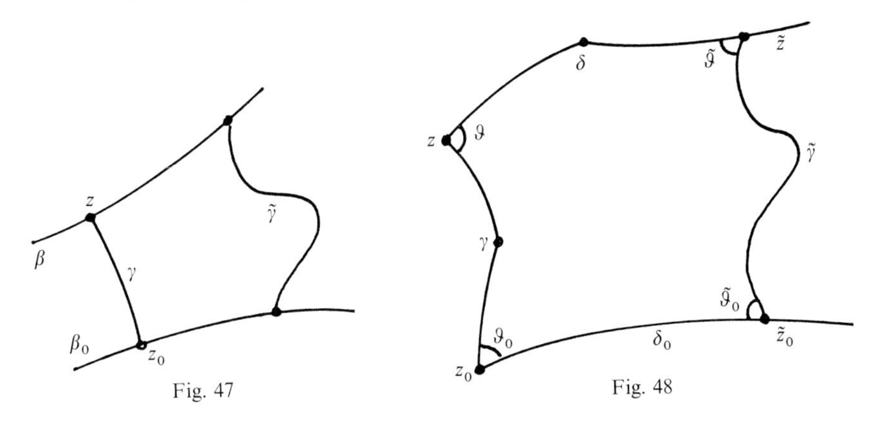

Fig. 47 Fig. 48

quadrilateral formed by the four arcs are smallest possible; in particular, $\vartheta_0 = \frac{\pi}{n_0+2}$, $\vartheta = \frac{\pi}{n+2}$.

Proof. The proof is, in principle, the same as for Theorem 16.1. We first construct a simple geodesic polygon Γ_0 containing both curves γ and $\tilde{\gamma}$ as well as the subarcs of δ_0 and δ between their points $z_0, \tilde{z}_0$ and $z, \tilde{z}$ respectively. It follows from Teichmüller's lemma (Theorem 14.1) that none of the vertical arcs β has a point in common with δ_0 or δ and therefore separates these two curves in G_0. We conclude, as before, that $|\tilde{\gamma}|_\varphi \geq |\gamma_\varphi$.

If equality holds, $\tilde{\gamma}$ must be a geodesic arc which traverses every strip S_{ij} exactly once, and orthogonally (i.e. parallel to the subarc γ_i of γ). Let us start with the first strip S_{11}. The only possibility that $\tilde{z}_0$ is on the boundary of S_{11} and that it is traversed orthogonally is that the angles ϑ_0 and $\tilde{\vartheta}_0$ are equal to $\pi/(n_0+2)$ and $\pi/(\tilde{n}_0+2)$ respectively and that the strip S_{11} follows δ_0, which means that all the respective angles are smallest possible. From the magnitude of ϑ_0 and $\tilde{\vartheta}_0$ we conclude, using Teichmüller's lemma again, that $\tilde{\gamma}$ and γ do not meet, unless $\tilde{z}_0 = z_0$, which we can exclude (we would get $\tilde{\gamma} = \gamma$ as in Theorem 16.2). The two curves γ and $\tilde{\gamma}$, together with the respective subarcs of δ_0 and δ, therefore form a quadrilateral Q. Continuing with the strip S_{12} we conclude that the line separating S_{11} and S_{12} cannot meet a zero of φ in Q. An argument similar to that for ϑ_0 shows that the interior angle at the endpoint of γ_1 must be smallest possible. Finally, a closer look at the possible zeroes of φ on $\tilde{\gamma}$ allows the same conclusion for the interior angles along this arc.

Conversely, if the angles $\vartheta_0, \ldots, \tilde{\vartheta}$ at the four corners of Q have the magnitude given in the theorem and the other angles on ∂Q are smallest possible (i.e. equal to $2\pi/(n_j+2)$ at a zero of order n_j), then Teichmüller's lemma applied to ∂Q shows, that φ has no zeroes in Q. Namely, the sum in formula (2) of Theorem 14.1 contains exactly four terms $\frac{1}{2}$, all the others are zero, hence $\sum n_i = 0$.

16.4 Generalization of the divergence principle. The inequality in the divergence principle can be slightly generalized. It is in fact not necessary to consider the entire arc $\tilde{\gamma}$, but only the subintervals of it lying in the strips orthogonal to γ. Already their length is $\geq |\gamma|_\varphi$.

Theorem 16.4. *Suppose γ is a geodesic arc of a holomorphic quadratic differential $\varphi \not\equiv 0$ in a simply connected region G. For simplicity we assume that the straight arcs α and α' through the endpoints of γ and cutting γ orthogonally are regular. The open strips S_i orthogonal to the straight subintervals γ_i of γ are disjoint (by the divergence principle). Let $\tilde{\gamma}$ be a rectifiable arc in G joining α to α'. Then, the total φ-length of the subarcs of $\tilde{\gamma}$ in $\bigcup S_i$ satisfies the length inequality*

$$|\tilde{\gamma} \cap (\bigcup S_i)|_\varphi \geq |\gamma|_\varphi.$$

Proof. Let Γ_0 be a simple φ-polygon in G which contains γ and $\tilde{\gamma}$ in its interior G_0. Since $\bar{G}_0$ contains only finitely many zeroes of φ, a construction similar to the one in the proof of Theorem 16.1 leads to finitely many open φ-rectangles S_{ij} in G_0, orthogonal to the sides γ_i of γ. Since $\tilde{\gamma}$ has its endpoints on α and α' respectively, it crosses every strip S_{ij}. We therefore have $|\tilde{\gamma}\cap S_{ij}|_\varphi \geq |\gamma_{ij}|_\varphi$, with $\gamma_{ij}=\gamma\cap S_{ij}$ the straight subinterval of γ in S_{ij}, crossing S_{ij} orthogonally. Summing up we get

$$\sum_{i,j}|\tilde{\gamma}\cap S_{ij}|_\varphi \geq \sum_{i,j}|\gamma_{ij}|_\varphi = |\gamma|_\varphi,$$

which proves the inequality.

§17. Minimal Length of Closed Geodesics

In this § we show that a closed trajectory, or, more generally, a closed geodesic, is a shortest curve in its free homotopy class. It is true for meromorphic quadratic differentials as well, if one considers homotopy with respect to the surface punctured at the poles of φ.

We first prove the theorem for a holomorphic quadratic differential in an annulus. Following Jenkins [3], we lift the differential to the universal covering surface and consider lifts of higher and higher powers of the closed curve, thus reducing the problem to the minimum length property of arcs (Section 16.1). In a second, more geometric proof, we apply the method of Section 16.1 to doubly connected domains. This approach allows for a generalization which is proved in Section 17.3.

17.1 Closed geodesics in an annulus

Theorem 17.1. *Let $\varphi \neq 0$ be a holomorphic quadratic differential in the annulus $\mathscr{A}$: $0 \leq r_0 < |z| < r_1 \leq \infty$, and let γ be a closed geodesic of φ. Then, any closed curve $\tilde{\gamma}$ in $\mathscr{A}$ which is freely homotopic to γ has φ-length $|\tilde{\gamma}|_\varphi \geq |\gamma|_\varphi$. (For the discussion of equality we refer to Section* 14.2.)

Proof. The geodesic γ is necessarily a Jordan curve separating the two boundary components of $\mathscr{A}$. It therefore has, in proper orientation, winding number one around $z=0$; the same is true for $\tilde{\gamma}$, since it is homotopic to γ. Choose $z\in\gamma$, $\tilde{z}\in\tilde{\gamma}$ and join the two points by an arc δ, $|\delta|_\varphi = d < \infty$. Lifting the quadratic differential and the curves to the parallel strip by the logarithm $w=\log z$, we get two arcs γ' and $\tilde{\gamma}'$ connecting consecutive lifts $w_i\in\gamma'$, $\tilde{w}_i\in\tilde{\gamma}'$, $i=1,2$ of z and $\tilde{z}$ respectively (Fig. 49). The lift δ' of δ connects w_1 and $\tilde{w}_1$. Of course, the lengths of an arc and its lift (with respect to the lifted differential) are the same. Considering n consecutive lifts we get a geodesic arc $n\cdot\gamma'$ the endpoints of which are joined by lifts of δ to the endpoints of the arc $n\cdot\tilde{\gamma}'$. By the minimum length property of geodesic arcs in simply connected domains (Theorem 16.1) we find

$$n\cdot|\gamma|_\varphi \leq n|\tilde{\gamma}|_\varphi + 2d,$$

hence, dividing by n and letting $n\to\infty$ $|\gamma|_\varphi \leq |\tilde{\gamma}|_\varphi$.

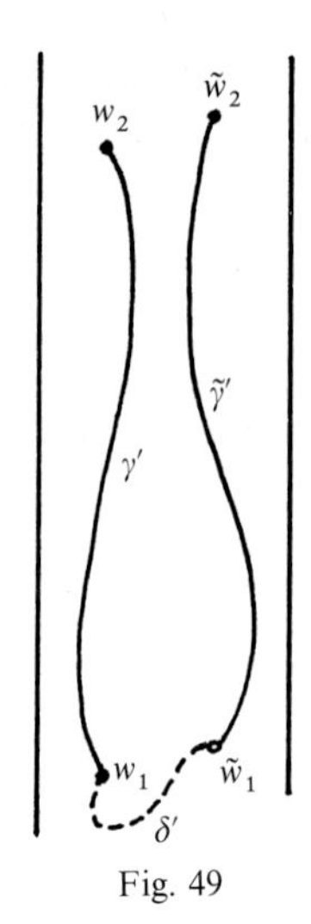

Fig. 49

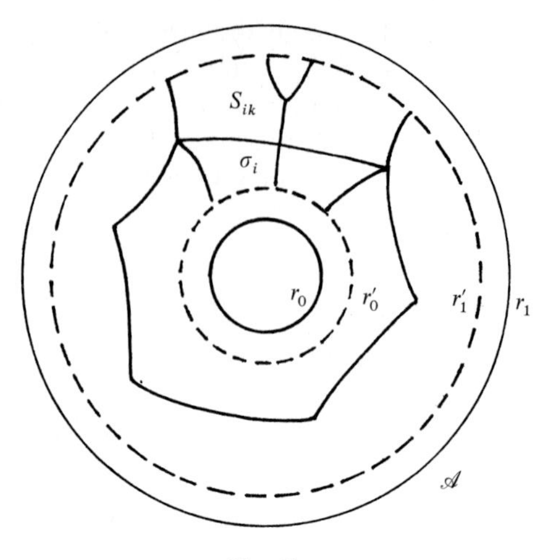

Fig. 50

17.2 Geometric proof of minimal length. A geometric proof of Theorem 17.1 can be given by lifting one circuit of γ to the parallel strip, as in the previous section, and then following Section 16.1 or rather 16.3. Although the lift of $\tilde{\gamma}$ does not connect the endpoints of the lift of γ, it still traverses all strips orthogonal to the lift of γ, exactly as in the proof of the divergence principle. Nevertheless, in view of the generalization in the next section, we give a direct construction for the annulus.

Theorem 17.2. *Suppose $\varphi \neq 0$ is a holomorphic quadratic differential in an annulus $\mathscr{A}$: $0 \leq r_0 < |z| < r_1 \leq \infty$, and γ is a closed geodesic of φ in $\mathscr{A}$. Let $\mathscr{A}'$: $r_0 < r_0' < |z| < r_1' < r_1$ be a relatively compact subring of $\mathscr{A}$. Then, there exists a finite set of open strips S_{ik} orthogonal to the sides σ_i of γ. Each strip S_{ik} is swept out by parallel θ_i-arcs orthogonal to σ_i and connects the two boundary circles of $\mathscr{A}'$* (Fig. 50).

Proof. The geodesic γ is a simple φ-polygon separating the two boundary circles of $\mathscr{A}$. We denote its sides by σ_i. If γ is not a closed (regular) θ-trajectory, φ has a zero at each vertex z_j of γ, of some order n_j. At these points both angles between the adjacent sides are $\geq \dfrac{2\pi}{n_j+2}$.

The annulus $\mathscr{A}'$ contains only finitely many zeroes of φ. Therefore there are only finitely many relatively critical arcs orthogonal to σ_i, for every i. An arc is called relatively critical (with respect to $\mathscr{A}'$) if it ends at a zero of φ, in at least one of its two directions, before hitting the boundary of $\mathscr{A}'$. All the other arcs orthogonal to σ_i are cross cuts of $\mathscr{A}'$, connecting its two boundary components, because no such arc can have two points in common with γ (this would violate the uniqueness of geodesic connections in simply connected domains). The domains S_{ik}, finitely many for each σ_i, swept out by these "radial" arcs, are the parallel strips we are looking for. No two arcs orthogonal to the same or different sides of γ can meet: this would violate the divergence principle. Therefore the strips S_{ik} are disjoint.

The boundary of each S_{ik} consists of geodesics parallel to S_{ik}, subintervals of the boundary circles of $\mathscr{A}'$ and, possibly, some subintervals of the continuations of the arcs sweeping out S_{ik}. The latter situation arises if such an arc ends up tangentially at a boundary circle of $\mathscr{A}'$. Since there are only finitely many such arcs, because of analyticity, a further subdivision would eliminate this possibility.

The length inequality is now easily proved. Suppose $\tilde{\gamma}$ is a closed curve in $\mathscr{A}$ freely homotopic to γ. We choose $\mathscr{A}'$ such that $\tilde{\gamma} \subset \mathscr{A}'$. The curve $\tilde{\gamma}$ passes through every strip S_{ik}: More accurately, it crosses each arc sweeping out S_{ik} at least once, for every i and k. Therefore its φ length is at least equal to the sum of the widths of the S_{ik}, which is $|\sigma_i|_\varphi$ for fixed i, and equal to $|\gamma|_\varphi = \sum_i |\sigma_i|_\varphi$ for the totality of S_{ik}.

17.3 Generalization to step curves (Strebel [8]). It is necessary, for many applications, to generalize the above result to closed curves which are composed of horizontal and vertical arcs in the way of circular segments and radial jumps. The reasoning of the previous section carries over without change.

Theorem 17.3. *Let $\varphi \neq 0$ be a holomorphic quadratic differential in the annulus $\mathscr{A}$: $0 \leq r_0 < |z| < r_1 \leq \infty$. Let $\gamma = \alpha_1 + \beta_1 + \alpha_2 + \beta_2 + \ldots + \alpha_n + \beta_n$ be a simple closed φ-polygon in $\mathscr{A}$ composed of horizontal arcs α_i and vertical arcs β_i. We assume that the two horizontal arcs α_i and α_{i+1} ($\alpha_{n+1} = \alpha_1$) which have points in common with β_i meet it from opposite sides (like a step). Then, any closed curve $\tilde{\gamma}$ which is freely homotopic to γ has φ-length $|\tilde{\gamma}|_\varphi \geq \sum_{i=1} a_i$, $a_i = |\alpha_i|_\varphi$.*

Proof. Teichmüller's lemma, applied to the interior of the curve γ, shows, that it must go around the disk $|z| \leq r_0$. The sum (4) of Theorem 14.1 now reads

$$\sum_j \left(1 - \frac{\vartheta_j}{\pi}\right) \geq 2$$

because all the n_j are equal to zero. The two angles adjacent to β_i are $\pi/2$ and $\frac{3\pi}{2}$, and therefore the two terms add up to zero; this gives a contradiction. The same argument shows that two vertical arcs cutting α_i and α_k respectively cannot meet: the terms coming from the angles adjacent to the β_i cancel. Then, there are three more angles, two of them being $\pi/2$. This shows, that the angle at the intersection of the two vertical arcs must be zero, hence they coincide.

We now perform the same construction as in the previous section, but only with the intervals α_i. The strips S_{ik}, orthogonal to the intervals α_i, are horizontal and not overlapping. The heights of the strips S_{ik} for fixed i add up to a_i. The curve $\tilde{\gamma}$, which necessarily crosses all the S_{ik}, therefore has length $|\tilde{\gamma}|_\varphi \geq \sum a_i$, as claimed.

17.4 Minimal length of closed geodesics on Riemann surfaces

Theorem 17.4. *Suppose $\varphi \neq 0$ is a holomorphic quadratic differential on an arbitrary Riemann surface R and γ is a closed geodesic of φ. Then, any closed curve $\tilde{\gamma}$ in the free homotopy class of γ has φ-length $|\tilde{\gamma}|_{\varphi} \geq |\gamma|_{\varphi}$.*

Proof. The curve γ cannot be homotopic to a point $P \in R$, as it would otherwise have a closed lift on the universal covering surface. We can therefore construct the annular covering surface $\hat{R}(\gamma)$ and thus reduce the problem to the previous one (Sections 17.1 or 17.2). If equality holds, $\tilde{\gamma}$ must be a geodesic (as it could otherwise be shortened in its free homotopy class). We now can apply Theorem 14.3.

§18. Existence of Geodesics

18.1 The basic existence theorem. Relatively shortest connection. In order to prove, under certain conditions, the existence of a geodesic arc joining two points in a simply connected domain (or in a given homotopy class on an arbitrary domain or on a Riemann surface), we prove the existence of a shortest curve connecting the two points. Our theorem is somewhat more general as it shows the existence of a unique shortest connection inside of a closed geodesic polygon.

Theorem 18.1. *Suppose D is the interior of a simple φ-polygon Γ. Let $\varphi \neq 0$ be holomorphic in $\bar{D}$, except for possible simple poles on Γ. Then, any two points z_1, $z_2 \in \bar{D}$ can be joined by a uniquely determined shortest line γ in $\bar{D}$* (Fig. 51).

Proof. Every point $z \in \bar{D}$ has a neighborhood $U(z)$ such that two arbitrary points $z', z'' \in U(z) \cap \bar{D}$ can be joined by a unique shortest arc in $U \cap \bar{D}$ (shortest

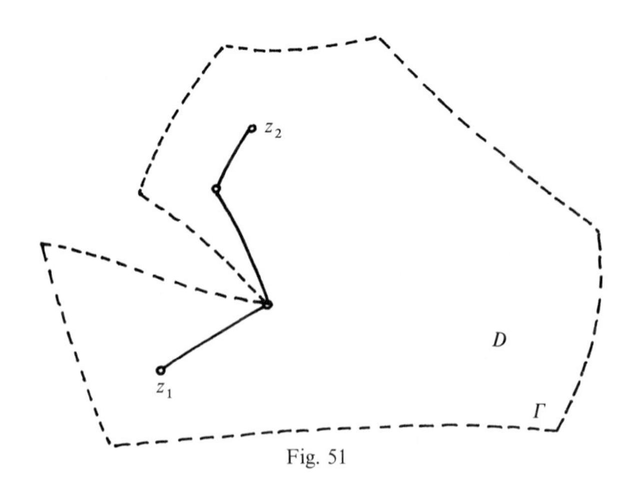

Fig. 51

in comparison with all connections in $\bar{D}$). This is clear for every point $z \in D$ (Theorem 8.1). The use of a distinguished local parameter in the neighborhood of a boundary point shows, that it is true there too. $\bar{D}$ can be covered by finitely many of these neighborhoods as it is compact in the φ-metric. If the interior distance of two points $z', z'' \in \bar{D}$ (the infimum of the φ-length of all arcs joining z' and z'' in $\bar{D}$) is small enough, there is a neighborhood in our covering which contains both of them. Hence they can be joined by a unique shortest arc in $\bar{D}$.

For an arbitrary arc γ joining z_1 to z_2 in $\bar{D}$ let $|\gamma|_\varphi$ be its φ-length. Denote by L the infimum of the φ-lengths of all such γ. L is clearly finite, and there is a minimal sequence (γ_n), $|\gamma_n|_\varphi \to L$. As parameter for γ_n we use the arc length in the φ-metric. We fix a number N and subdivide the parameter interval $[0, |\gamma_n|_\varphi]$ into N equal parts. N is chosen so that two neighboring points on γ_n can be joined by a unique shortest arc in $\bar{D}$. For a properly chosen subsequence (γ_{n_i}) the $N+1$ distinguished points converge. If we join the limits by respective shortest arcs in $\bar{D}$ we have found our shortest curve γ. It clearly is a geodesic arc with respect to φ, except possibly at the boundary points of D, ∂D serving as obstacle.

To prove uniqueness, let $\tilde{\gamma}$ be an arbitrary locally shortest connection between z_1 and z_2 in $\bar{D}$. If the two curves do not coincide they contain subarcs which, together, form a closed Jordan curve. Without loss of generality we can assume that this is the case for the original curves. At every vertex of the Jordan curve $\gamma \cup \tilde{\gamma}$, except for the two points z_1, z_2 (where the angles are still positive), the angle condition is fulfilled. This is evident for the points in D. But a closer look shows that it is also true for the interior angle at a possible vertex on ∂D, as the obstacle pushes the curve towards the interior and therefore opens up the respective angle. An application of Teichmüller's lemma to the curve $\gamma \cup \tilde{\gamma}$ gives a contradiction.

18.2 Existence of a shortest connection. The main application of Theorem 17.1 is to show the existence of a geodesic arc between two given points.

Theorem 18.2.1. *Let $\varphi \not\equiv 0$ be a holomorphic quadratic differential in a simply connected domain D, $z_1, z_2 \in D$. Denote by $d_\varphi(z_1, z_2)$ the φ-distance of the two given points $z_1, z_2 \in D$, i.e. the infimum of the φ-lengths of all arcs joining z_1 to z_2 in D. If the φ-distance of z_1 from the boundary of D is greater than $d_\varphi(z_1, z_2)$, then there exists a shortest (and hence geodesic) arc joining z_1 to z_2 in D. By Theorem* 14.2.1 *it is uniquely determined.*

Proof. We can choose a Jordan subdomain D_0 which is bounded by a simple closed φ-polygon $\Gamma_0 = \partial D_0$ and such that $d_\varphi(z_1, \Gamma_0) > d_\varphi(z_1, z_2)$: we start with a Jordan domain which satisfies this condition and then replace sufficiently short subintervals of its boundary by geodesic arcs. Applying Theorem 17.1 to D_0 gives the desired result. The uniqueness Theorem 14.2.1 shows that it does not depend on D_0.

The result is true for arbitrary points $z_1, z_2 \in D$, if ∂D is at infinite distance i.e. the φ-metric is complete. We have

Corollary 18.2. *Let D be a simply connected domain whose boundary is at infinite φ-distance, where φ is holomorphic in D. Then any two points of D can be joined by a unique shortest line (and hence a uniquely determined geodesic).*

Lifts of quadratic differentials of compact surfaces have the above property. We state this as a

Lemma 18.2 (Ahlfors [2]). *Suppose $\varphi \neq 0$ is a holomorphic quadratic differential on a compact Riemann surface R. Then, the lift $\hat{\varphi}$ of φ to the universal covering surface $\hat{R}$ of R induces a complete metric.*

Proof. It is evident if R is the torus, for then $\hat{R}$ is the plane and $\hat{\varphi}=$const.

Let the genus of R be greater than one and represent $\hat{R}$ as the disk $|z|<1$. Pick two numbers $0<r_0<r_1<1$ such that the disk $|z|<r_0$ covers a fundamental region of the group G of cover transformations of $\hat{R}$. Let $d>0$ be the $\hat{\varphi}$-distance of the two circles $|z|=r_0$ and $|z|=r_1$. Every point z on $|z|=r_1$ has an equivalent point in $|z|<r_0$ under the group G. By compactness, the circle $|z|=r_1$ can be covered by finitely many images of the disk $|z|<r_0$ under elements of G. Look at the images of the disk $|z|<r_1$ by these transformations. If $r_2<1$ is chosen such that all these disks are contained in the disk $|z|<r_2$, then the $\hat{\varphi}$-distance of $|z|=r_2$ from $z=0$ is at least $2d$. The argument can be continued indefinitely. This proves the lemma.

The usual lifting procedure now leads to the following

Theorem 18.2.2. *Let γ be an arc joining two given points P_1 and P_2 of a compact Riemann surface R and let $\varphi \not\equiv 0$ be a holomorphic quadratic differential on R. Then there is a unique shortest curve γ_0 with respect to the φ-metric in the homotopy class of γ.*

18.3 Minimal length of geodesic arcs. Theorem 18.1 furnishes another proof of the minimum length property of geodesic arcs. Let φ be holomorphic in a simply connected plane domain D. Let γ be a geodesic arc and $\tilde{\gamma} \neq \gamma$ an arbitrary arc joining two points $z_1, z_2 \in D$. As in the proof of Theorem 18.1 we choose a subdomain D_0 which contains both curves. Let γ_0 be the shortest curve in $\bar{D}_0$ connecting the two points. The uniqueness part of Theorem 18.1 shows that $|\gamma_0|=|\gamma|$. Therefore $|\tilde{\gamma}|>|\gamma|$.

18.4 Existence of closed geodesics on compact Riemann surfaces. The reasoning which served to prove Lemma 18.2 can also be used to show the following

Lemma 18.4. *Suppose $\varphi \neq 0$ is a holomorphic quadratic differential on a compact Riemann surface R of genus ≥ 2. Let $\hat{\varphi}$ be its lift to the universal covering surface D of R. Then, any arc γ in D which spans a fixed angle becomes arbitrarily long if it is sufficiently close to ∂D.*

Proof. Let σ_0 and σ be two radii in $\hat{R}=D$ spanning a sector Δ with angle ϑ at its vertex zero. Subdivide Δ by radii σ_i, $i=0, \ldots, n$, $\sigma_n=\sigma$, forming equal angles of magnitude ϑ/n. Using the notations of Lemma 18.2, every point of σ_i,

$i=0, \ldots, n$, has an equivalent point in the disk $|z|<r_0$ under the group of cover transformations of $\hat{R}$. We can clearly choose $r<1$ such that, for any points $z_i \in \sigma_i$, $r<|z_i|<1$, the images of the larger disk $|z|<r_1$ under the corresponding cover transformations, are disjoint. Suppose γ is an arc joining σ_0 to σ and crossing Δ in the annulus $r<|z|<1$. Then, its $\hat{\varphi}$-length is greater than $n \cdot d$, and since, for $r \to 1$, the number n can be made arbitrarily large, $|\gamma|_{\hat{\varphi}}$ becomes arbitrarily large.

Theorem 18.4. *Let $\varphi \neq 0$ be a holomorphic quadratic differential on a compact Riemann surface R of genus ≥ 2. For every loop γ on R which is not retractible to a point there exists a shortest curve γ_0 freely homotopic to γ.*

Proof. Let $\hat{R}(\gamma)$ be the annular covering surface associated with γ and let $\hat{\gamma}$ be a closed lift of γ to $\hat{R}(\gamma)$: denote the infimum of the $\hat{\varphi}$-lengths of all simple loops separating the two boundary components of $\hat{R}(\gamma)$ by L, and choose a minimizing sequence $(\hat{\gamma}_n)$,

$$L = \lim_{n \to \infty} |\hat{\gamma}_n|_{\hat{\varphi}}.$$

Pick a point $z_n \in \hat{\gamma}_n$, for every n. Because of Lemma 18.2 and Lemma 18.4, the points z_n are bounded away from the boundary of $\hat{R}(\gamma)$. We can assume, by passing to a subsequence, that the sequence (z_n) converges to a point $z_0 \in \hat{R}(\gamma)$. An application of Corollary 18.2 to the simply connected covering surface of $\hat{R}(\gamma)$ shows that there exists a simple loop $\hat{\gamma}_0$ in $\hat{R}(\gamma)$ starting at z_0 and returning to it. It is a geodesic except possibly at z_0. Going from z_0 along a straight arc to z_n, then around $\hat{\gamma}_n$ and back to z_0 we see that

$$L \leq |\hat{\gamma}_0|_{\hat{\varphi}} \leq \lim_{n \to \infty} |\hat{\gamma}_n|_{\hat{\varphi}} = L.$$

Therefore $\hat{\gamma}_0$ has minimal length among the simple loops separating the boundary components of $\hat{R}(\gamma)$. It is a geodesic also at z_0, because it could be shortened otherwise. The projection γ_0 of $\hat{\gamma}_0$ on R is the geodesic in the free homotopy class of γ.

§ 19. Holomorphic Quadratic Differentials of Finite Norm in the Disk

19.1 Horizontal strips. The family of holomorphic quadratic differentials φ of finite norm $\|\varphi\| = \iint_{|z|<1} |\varphi(z)|\, dx\, dy < \infty$ in the unit disk D is nothing else than the Banach space of integrable holomorphic functions in D. If we consider, instead of the unit disk, an arbitrary simply connected domain G, it must be bounded by a continuum, unless $\varphi \equiv 0$ (see Sections 5.3 and 12.2). It is therefore conformally equivalent with D, and this more general case can be reduced to ours, using the transformation law for quadratic differentials.

In this first section, the finiteness of the norm is however not needed. We have already seen (Section 15.1) that every maximal geodesic arc, in particular every non critical trajectory, is a cross cut of D, which means that the limit sets of both of its rays are on the boundary $|z|=1$. We now consider strips of parallel trajectories.

Definition 19.1. Let φ be a holomorphic quadratic differential in a simply connected domain G and let β be an open vertical arc (not necessarily maximal) of φ. The subset of G covered by the horizontal trajectories through the points of β is called the horizontal strip $S=S(\beta)$ determined by β. More generally, if $E\subset\beta$ is any set, the trajectories through the points of E form the horizontal strip $S(E)$.

We choose a branch of Φ on β and continue its inverse Φ^{-1} conformally, as far as possible, along the horizontals through the points $w\in\beta'=\Phi(\beta)$ (see Section 13.2). The set of points covered by these maximal open horizontal intervals $\alpha'=\alpha'(w)$ is a simply connected domain S' which we call the Euclidean horizontal strip determined by β and the chosen branch of Φ. Likewise we speak of the Euclidean horizontal strip determined by $E'=\Phi(E)$.

The simple connectivity of $S'(\beta)$ follows from the fact that with any $w'\in S'$ the interval $[w,w']$ on the horizontal through w' with one endpoint w', the other $w\in\beta'$, is contained in S'. Therefore any closed curve $\gamma'\subset S'$ can be contracted, in S', to β' and then to a point.

Theorem 19.1. *Let S' be the Euclidean horizontal strip determined by the open vertical arc β and the chosen branch Φ. Then the inverse function Φ^{-1} is a univalent conformal mapping of S' onto S.*

Proof. (See also Section 13.2.) The univalence is a consequence of the simple connectivity of G. For, let $w_1, w_2\in S'$, $\Phi^{-1}(w_1)=\Phi^{-1}(w_2)=z$ (Fig. 52). Then, the

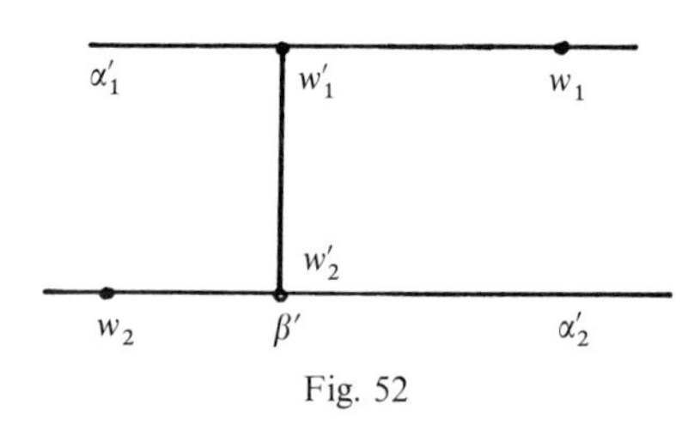

Fig. 52

two horizontal intervals $\alpha_k'=S'\cap\{w\,|\,\operatorname{Im} w=\operatorname{Im} w_k\}$, $k=1,2$, are mapped by Φ^{-1} onto the trajectory α through z. Let $w_k'=\alpha_k'\cap\beta'$. If $w_1'\neq w_2'$, β is either closed (if $\Phi^{-1}(w_1')=\Phi^{-1}(w_2')$) or else it has two different points in common with α (if $\Phi^{-1}(w_1')\neq\Phi^{-1}(w_2')$), an impossibility. Therefore $w_1'=w_2'$, and hence $\alpha_1'=\alpha_2'$. Again, if $w_1\neq w_2$, α is closed, which is impossible, and thus $w_1=w_2$.

19.2 Coverings of G by horizontal strips (Strebel [2]). For many applications it is useful to have an exhaustion of G by at most denumerably many horizontal strips. Let S_1 and S_2 be two open horizontal strips determined by the vertical intervals β_1 and β_2 respectively. Then, with any $z\in S_1\cap S_2$ the entire trajectory α through z is contained in both strips. Therefore $S_1\cap S_2$ is a union of

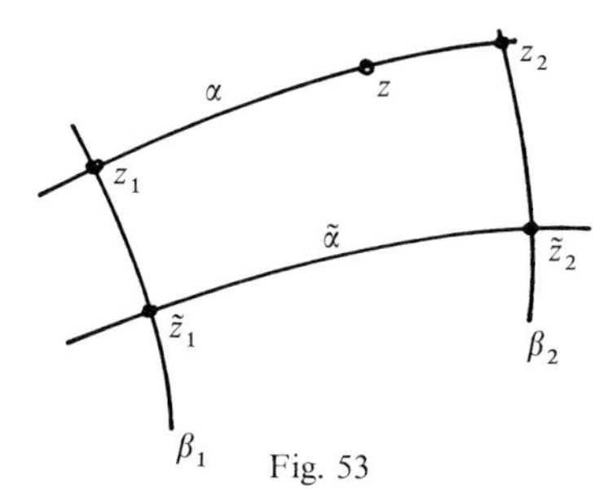

Fig. 53

trajectories. Moreover, every trajectory α which passes through a certain neighborhood of z is also contained in $S_1 \cap S_2$. Consider two trajectories α and $\tilde{\alpha}$ in $S_1 \cap S_2$ (Fig. 53). Let their intersections with β_1 and β_2 be z_1, z_2 and $\tilde{z}_1, \tilde{z}_2$ respectively. Then, by the divergence principle (Theorem 16.3), the φ-length of the subarc $[z_2, \tilde{z}_2]$ of β_2 is at least equal to the φ-length of the arc $[z_1, \tilde{z}_1]$ of β_1, and conversely. We therefore have equality: $\|[z_1, \tilde{z}_1]\|_\varphi = \|[z_2, \tilde{z}_2]\|_\varphi$ and thus the rectangle with the vertices $z_1, \tilde{z}_1, \tilde{z}_2, z_2$ and the respective subintervals of α, β_1, $\tilde{\alpha}$ and β_2 as sides does not contain any zeroes of φ. We conclude that every trajectory through a point $z \in [z_1, \tilde{z}_1]$ of β_1 is contained in $S_1 \cap S_2$. Therefore the intersection $S_1 \cap S_2$ consists of the trajectories through an open subinterval of β_1 (and of β_2). The difference $S_1 \setminus S_2$ is the set of trajectories or, more accurately, the set of points of G covered by these trajectories, which pass through at most two half open subintervals of β_1; the same holds for β_2. We can now prove the following

Theorem 19.2. *Let φ be a holomorphic quadratic differential in a simply connected domain G. Then, there is a denumerable sequence of horizontal strips S_k, determined by open, half open or closed vertical intervals β_k, which cover G up to the zeroes of φ. (If we choose the β's open, the covering is by open strips and up to a denumerable set of horizontal trajectories and points.)*

Proof. Let (z_ν) be a denumerable set of regular points of φ which is dense in G. Denote by S_ν the open horizontal strip determined by the vertical trajectory β_ν through z_ν. If $z \in G$ is a regular point of φ, it is the center of a φ-square Q. For every $z_\nu \in Q$, the vertical trajectory β_ν traverses Q and therefore $z \in S_\nu$, which shows that the union $\bigcup S_\nu$ covers G up to the zeroes of φ. We now form the following sequence:

$$S_1, S_2 \setminus S_1, S_3 \setminus S_2 \setminus S_1, \ldots, S_\nu \setminus S_{\nu-1} \setminus \ldots \setminus S_1, \ldots.$$

The general term

$$S_\nu \setminus S_{\nu-1} \setminus S_{\nu-2} \setminus \ldots \setminus S_1$$

consists of finitely many horizontal strips determined by open, half open or closed subintervals of β_ν. The union of the above strips is therefore the desired covering of G. If we leave possible endpoints of the vertical intervals away, we have an open covering of G (up to a set of measure zero).

19.3 The φ-length and the Euclidean length of the trajectories. From now on we assume that φ has finite norm. We can then take G to be the unit disk D. Let $S=S(\beta)$ be the horizontal strip determined by the open vertical interval β. As Φ^{-1} maps the corresponding Euclidean strip S' homeomorphically onto S, we have (see also Section 13.2)

$$|S|_\varphi=\iint_S |\varphi(z)|\,dx\,dy=\iint_{S'} du\,dv=|S'|,$$

and of course

$$|\alpha|_\varphi=\int_\alpha |\varphi(z)|^{1/2}\,|dz|=\int_{\alpha'} du=|\alpha'|.$$

By $|S'|$ and $|\alpha'|$ we denote Euclidean area and length respectively. But the representation of the quadratic differential in terms of the parameter w is $\varphi\equiv 1$. Therefore the two identities are in a way trivial: They just say that φ-area and φ-length are invariantly defined.

As $|S|_\varphi\leq\|\varphi\|<\infty$, the Euclidean area $|S'|$ of S' is finite, and therefore the length of the horizontal $\alpha'=\alpha'(v)$ in S' is finite for a.a. v. We conclude, that the trajectories through a.a. points of β have finite φ-length.

The same statement can be made about the Euclidean length of the trajectories. Let α be the trajectory through $z\in\beta$, $w=\Phi(z)=u+iv$. The Euclidean length of $\alpha=\alpha(v)$ is equal to

$$|\alpha(v)|=\int_\alpha |dz|=\int_{\alpha'}\left|\frac{dz}{dw}\right| du.$$

Integration over v and squaring gives

$$\left(\iint_{S'}\left|\frac{dz}{dw}\right| du\,dv\right)^2\leq\iint_{S'} du\,dv\cdot\iint_{S'}\left|\frac{dz}{dw}\right|^2 du\,dv\leq|S'|\cdot\pi\leq\|\varphi\|\cdot\pi,$$

which proves the assertion.

Corollary 19.3. *Let φ be a holomorphic quadratic differential of finite norm in the unit disk. Then, a.a. horizontal trajectories have finite Euclidean length (i.e. the set of horizontal trajectories of infinite length covers a set of two dimensional measure zero).*

Proof. It is true in every horizontal strip, and the entire disk can be covered by denumerably many horizontal strips.

19.4 The boundary behaviour of the trajectories. If α is a non critical trajectory of finite Euclidean length, then its two rays α^+ and α^- converge to well determined boundary points ζ^+ and ζ^- respectively. We now want to show that the two limits are different. More generally, we prove

Theorem 19.4. *Let φ be a holomorphic quadratic differential of finite norm in the unit disk D. Then*

a) *the cluster sets A^+ and A^- of the two rays α^+ and α^- of a non critical trajectory α on ∂D are disjoint; more generally;*

b) *the cluster sets* A_1^+ *and* A_2^+ *of two rays* α_1^+ *and* α_2^+ *with initial points* z_1 *and* z_2, *respectively, on the same geodesic* γ *of* φ, *are disjoint;*

c) *the cluster sets* A^+ *and* B^+ *of a trajectory ray* α^+ *and a vertical trajectory ray* β^+ *with the same initial point* z_0 *are disjoint. (In Section 19.6 it will be shown that these cluster sets in fact always consist of a single point.)*

Proof. To show a), let $\zeta^+ \in A^+$ and $\zeta^- \in A^-$ and assume $\zeta^+ = \zeta^- = \zeta$ (Fig. 54). We fix a compact subinterval α_0 of α. Let $d>0$ be its φ-length. We consider

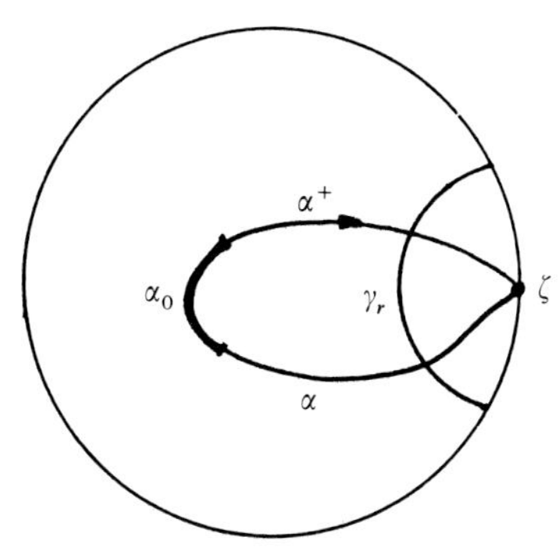

Fig. 54

the segment γ_r of the circle $|z-\zeta|=r$ which is in D, for all $r>0$, for which α_0 stays outside of the disk D_r: $|z-\zeta|\le r$. As any horizontal interval is the shortest connection between its two endpoints and as, by assumption, γ_r meets both rays α^+ and α^-, $|\gamma_r|_\varphi \ge |\alpha_0|_\varphi = d$. Therefore

$$d \le \int_{\gamma_r} |\varphi(r e^{i\vartheta})|^{1/2}\, r\, d\vartheta.$$

An application of the Schwarz inequality yields

$$\int_0^\rho \frac{d^2}{r}\, dr \le \iint_{D\cap D_\rho} |\varphi(r e^{i\vartheta})|\, r\, dr\, d\vartheta \le \|\varphi\| < \infty,$$

which is impossible.

For the proof of b) we use, instead of α_0, the subinterval of γ with endpoints z_1 and z_2 and apply the divergence principle (Section 16.3). Finally, to prove c), we use a compact subinterval β_0 of β^+ and again apply the divergence principle.

It should be noticed that the statement a) is still true if only the restriction of φ to the interior G of α has finite norm, assuming that the intersections of all the arcs γ_r with G span a certain positive angle centered at ζ. Similar remarks can be made for b) and c).

Corollary 19.4. *Let* φ *be holomorphic and of finite norm in the disk* D. *Then, at most denumerably many trajectory rays can diverge* (see also Section 19.6).

Proof. The statement is true for the rays with initial point on the same vertical interval β, as the cluster set of any ray is a closed subinterval of ∂D

and all the subintervals are disjoint. As we can choose denumerably many vertical intervals such that their union cuts every horizontal trajectory, the statement follows.

19.5 The length inequality for cross cuts. In this section we want so show the length inequality for convergent cross cuts.

Theorem 19.5. *Let φ be a holomorphic quadratic differential of finite norm in the disk D and let γ be a geodesic arc both ends of which converge to points ζ_1 and ζ_2 on ∂D respectively. Then any arc $\tilde{\gamma}$ in D joining ζ_1 and ζ_2 has φ-length $|\tilde{\gamma}|_\varphi \geq |\gamma|_\varphi = c$. If c is finite, equality holds only for $\tilde{\gamma} = \gamma$.*

Proof (Fig. 55). Let $\gamma_r^{(i)}$ be the intersection of the circle $|z-\zeta_i| = r$ with D, $i = 1, 2$. Denoting its φ-length by $\ell_i(r)$ we get

$$\ell_i(r) = \int_{\gamma_r^{(i)}} |\varphi(z)|^{1/2} |dz| = \int_{\gamma_r^{(i)}} |\varphi(r e^{i\vartheta})|^{1/2} r \, d\vartheta.$$

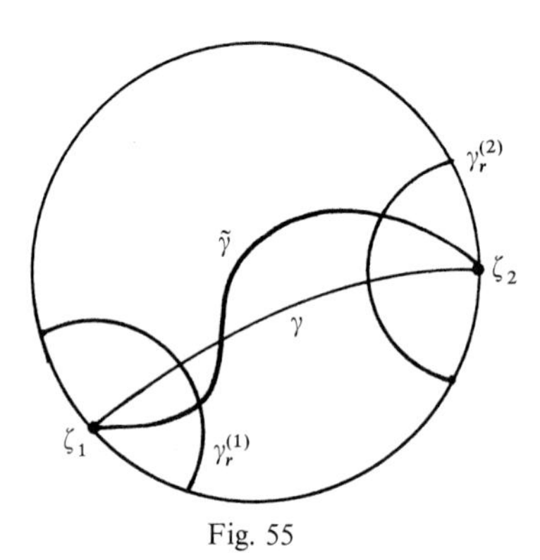

Fig. 55

Squaring and using the Schwarz inequality yields

$$\ell_i^2(r) \leq r\pi \int_{\gamma_r^{(i)}} |\varphi(r e^{i\vartheta})| \, r \, d\vartheta,$$

and after integration

$$\int_0^{r_0} \frac{\ell_i^2(r)}{r} dr \leq \pi \iint_{D_{r_0}^{(i)} \cap D} |\varphi(z)| \, r \, dr \, d\vartheta \leq \pi \|\varphi\|.$$

We conclude that there is a sequence of radii, tending to zero, for which the φ-length ℓ_i tends to zero, $i = 1, 2$. As $\gamma_r^{(i)}$ cuts γ and $\tilde{\gamma}$ for every r, there exist subarcs of $\tilde{\gamma}$ of φ-length greater than $c - 2\varepsilon$, for every $c < |\gamma|_\varphi$ and every $\varepsilon > 0$. This proves the inequality $|\tilde{\gamma}|_\varphi \geq |\gamma|_\varphi$. Let now $|\gamma|_\varphi < \infty$ and let equality hold. Then $\tilde{\gamma}$ must be a geodesic, as it could otherwise be shortened to give $|\tilde{\gamma}|_\varphi < |\gamma|_\varphi$. From the uniqueness of geodesic connections we conclude that $\tilde{\gamma} \cap \gamma$ is connected (either empty or a single point or a single interval). If $\tilde{\gamma} \neq \gamma$ we therefore find a point $z \in \gamma$ and a small vertical interval β (Fig. 56) with end-points z and z' and such that z' lies on different sides of γ and $\tilde{\gamma}$ (if we orient γ

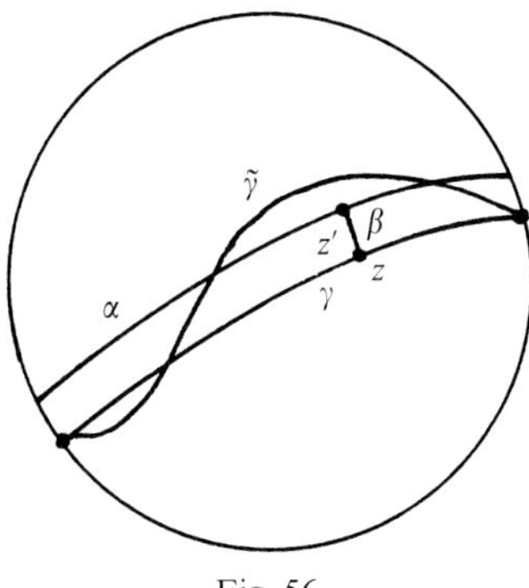

Fig. 56

and $\tilde{\gamma}$ in the same way, z' lies on the left of γ and on the right of $\tilde{\gamma}$, e.g.). We can, of course, choose z' such that it is a point of a non critical Θ-trajectory α, parallel to γ at z, with convergent ends. Because of the divergence principle γ and α have different endpoints. Therefore there is a subinterval of $\tilde{\gamma}$ which has exactly its two endpoints on α, contradicting the uniqueness of geodesic connections. This proves the theorem.

19.6 The boundary behaviour of the trajectories (continued). In this section we show that every geodesic ray of a holomorphic quadratic differential of finite norm in the unit disk converges to a well determined boundary point.

Let $\varphi \not\equiv 0$ be holomorphic in the open disk D with $\|\varphi\| < \infty$. The φ-distance of two points z_1, and z_2 in D is defined to be

$$d(z_1, z_2) = \inf_{\{\gamma\}} |\gamma|_\varphi = \inf_{\{\gamma\}} \int_\gamma |\varphi(z)|^{1/2} \, |dz|,$$

where γ ranges over the piecewise smooth curves in D joining z_1, to z_2. If z_2 lies on the boundary of D, the distance $d(z_1, z_2)$ is defined to be

$$d(z_1, z_2) = \liminf_{z_2^{(n)} \to z_2} d(z_1, z_2^{(n)})$$

for all sequences $(z_2^{(n)})$ of points of D tending to z_2. Equivalently,

$$d(z_1, z_2) = \lim_{r \to 0} \inf \{d(z_1, z)|; |z - z_2| < r, \ z \in D\}.$$

If both points z_1 and z_2 are boundary points of D, the generalization of the definition is achieved by simultaneous approximation of z_1 and z_2 from the interior.

The function d is non negative and clearly satisfies the triangle inequality. It is also evident that $d(z_1, z_2) > 0$ if at least one of the two points z_1 and z_2 is an interior point of D. For then the approximating arcs have to leave a certain fixed neighborhood of that interior point and thus their lengths have a positive lower bound. We are now going to show that the same is true for two boundary points of D.

Lemma 19.6. *Let $\varphi \not\equiv 0$ be a holomorphic quadratic differential of finite norm in the disk D. Then the φ-distance $d(\zeta_1, \zeta_2)$ of any two boundary points ζ_1 and $\zeta_2 \neq \zeta_1$ is positive.*

Proof. Choose $0<\rho<1$. The φ-length of the radial interval $\arg z = \vartheta$, $\rho \leq |z| < 1$ is

$$\ell_\rho(\vartheta) = \int_\rho^1 |\varphi(r\,e^{i\vartheta})|^{1/2}\,dr.$$

As $\rho \leq r$, we get

$$\int_0^{2\pi} \ell_\rho(\vartheta)\,d\vartheta \leq \frac{1}{\rho}\int_0^{2\pi}\int_\rho^1 |\varphi(r\,e^{i\vartheta})|^{1/2}\,r\,dr\,d\vartheta,$$

and the Schwarz inequality then yields

$$\left(\int_0^{2\pi} \ell_\rho(\vartheta)\,d\vartheta\right)^2 \leq \pi\,\frac{1-\rho^2}{\rho^2}\int_0^{2\pi}\int_\rho^1 |\varphi(r\,e^{i\vartheta})|\,r\,dr\,d\vartheta < \infty.$$

We conclude that a.a. radii have finite φ-length. This argument also shows that a.a. points of the boundary are at finite distance from the center, or from any other point $z \in D$.

We now assume that there are two points ζ_1 and $\zeta_2 \neq \zeta_1$, on ∂D with $d(\zeta_1, \zeta_2) = 0$ (Fig. 57). Then there is a sequence of arcs γ_n with endpoints $z_1^{(n)}$ and

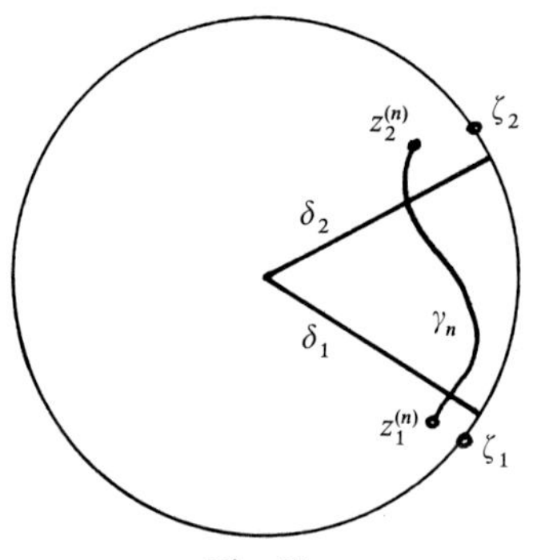

Fig. 57

$z_2^{(n)}$ tending to ζ_1 and ζ_2 respectively and such that $|\gamma_n|_\varphi \to 0$. Clearly the sequence (γ_n) tends to ∂D. For otherwise there would exist an annulus $\rho_1 < r < \rho_2$ in D which is crossed twice by γ_n for every sufficiently large n, and therefore twice the φ-distance of $|z| = \rho_1$ and $|z| = \rho_2$ would be a lower bound for $|\gamma_n|_\varphi$. At least one of the intervals on ∂D between ζ_1 and ζ_2, Γ_0, say, is approximated by the sequence (γ_n). We choose points ζ_1' and $\zeta_2' \neq \zeta_1'$ in this interval such that the corresponding radii δ_i: $\arg z = \arg \zeta_i'$, $i = 1, 2$, have finite φ-length. Let G_n be the domain bounded by subintervals of δ_1, δ_2 and a subarc of γ_n joining these two radii. As

$$\int_{\partial G_n} |\varphi(z)|^{1/2}\,|dz|$$

is uniformly bounded, φ belongs to the Hardy class $H^{1/2}$ of the sector G bounded by the two radii δ_i and the subinterval Γ_0 of ∂D which joins them

(Duren [1], p. 168). A function of any Hardy class in G has an angular limit at almost every point of the boundary. Let this limit be different from zero on a subset E of Γ_0 of positive measure. Then, by Egoroff's theorem, there is a set $E_0 \subset E$ of positive measure, a number $\varepsilon > 0$ and a radius $\rho < 1$ such that for $\vartheta \in E_0$, $\rho < r < 1$, $|\varphi(re^{i\vartheta})| > \varepsilon^2$. We conclude that $|\gamma_n|_\varphi > \rho \cdot \varepsilon \cdot |E_0|$ for all sufficiently large n, which is a contradiction. Therefore the radial limit of φ is equal to zero a.e. on Γ_0, and since $\varphi \in H^{1/2}(G)$, $\varphi \equiv 0$, finishing the proof of the lemma (Duren [1]).

It was remarked by Li Zhong that the above argument goes through for quadratic differentials φ with the following property: There is a dense set of points ζ, $|\zeta| = 1$, which can be joined to $z = 0$ by Jordan arcs of finite φ-length.

We are now ready to prove the convergence of geodesic rays.

Theorem 19.6. *Let $\varphi \not\equiv 0$ be a holomorphic quadratic differential of finite norm in the disk D or, more generally, a quadratic differential with the above finiteness property. Then, every maximal geodesic ray γ^+ converges to a well determined boundary point of ∂D.*

Proof. Let γ^+ be a geodesic ray of φ with initial point $z_0 \in D$ which has more than one accumulation point. Since γ^+ tends to the boundary of D, its set of accumulation points is a closed, non degenerate interval $I \subset \partial D$. Let ζ be a point in the interior of I which can be joined to z_0 by an arc δ of finite φ-length. By assumption, δ lies in D, except for its endpoint ζ. Since the arc γ^+ approximates I, there is a sequence of points $z_n \in \gamma^+ \cap \delta$ with the property that the subarcs $\gamma^+(z_0, z_n)$ of γ^+ with initial and endpoints z_0 and z_n respectively exhaust γ^+. By the length inequality for geodesic arcs

$$|\gamma^+(z_0, z_n)|_\varphi \leq |\delta(z_0, z_n)|_\varphi \leq |\delta|_\varphi,$$

with $\delta(z_0, z_n)$ the subarc of δ between z_0 and z_n. Therefore

$$|\gamma^+|_\varphi \leq |\delta|_\varphi < \infty.$$

Let now ζ_1 and $\zeta_2 \neq \zeta_1$ be two points in the interior of I which can be joined to z_0 by arcs δ_1 and δ_2 respectively of finite φ-length. We may clearly assume that δ_1 and δ_2 only have z_0 in common. There exists a sequence of disjoint subarcs $\gamma^{(n)}$ of γ^+ joining points $z_1^{(n)} \in \delta_1$ and $z_2^{(n)} \in \delta_2$ and tending to I. Since $\sum |\gamma^{(n)}|_\varphi \leq |\gamma^+|_\varphi$, $|\gamma^{(n)}|_\varphi \to 0$. Lemma 19.6 leads to a contradiction. This proves the theorem.

19.7 Quadratic differentials in the disk, arising from quadratic differentials of finite norm on a Riemann surface. Suppose R is a Riemann surface the universal covering surface $\hat{R}$ of which is represented by the disk D. A holomorphic quadratic differential φ on R has a uniquely determined lift $\hat{\varphi}$ on $\hat{R}$. Although $\hat{\varphi}$ does not have finite norm, even if φ does (except for simply connected R), the trajectory structure of $\hat{\varphi}$ is topologically the same as for a $\hat{\varphi}$ with $\|\hat{\varphi}\| < \infty$. The following facts can easily be proved, by a combination of the arguments of the preceding sections with considerations about Fuchsian groups (for details see A. Marden and K. Strebel [2]):

(1) Every geodesic ray $\hat{\gamma}$ of $\hat{\varphi}$ tends to a well determined boundary point. (Of course, the ray $\hat{\gamma}$ is assumed to be maximal, i.e. it does not stop at any point in $\hat{R}$, whether regular nor critical, but is continued indefinitely.) If R is of the first kind, i.e. its limit set Λ is the entire boundary ∂D, the assumption of finite norm is not needed.

(2) Let γ_1 and γ_2 be two geodesic rays on R with the same initial point 0 but starting in different directions (with respect to a local parameter near 0). Then, the two lifts $\hat{\gamma}_1$ and $\hat{\gamma}_2$ of γ_1 and γ_2 respectively with the same initial point $\hat{0}$ above 0 have different endpoints. In particular, the two endpoints of a geodesic on ∂D are different.

(3) Assume that γ_1 and γ_2 are geodesics which have lifts $\hat{\gamma}_1$ and $\hat{\gamma}_2$ respectively with the same endpoints on ∂D. Then γ_1 and γ_2 are closed. They bound an annulus on R swept out by closed, regular θ-trajectories (i.e. lines $\arg \varphi(z)\, dz^2 = \theta$).

19.8 Generalization. The metric $|\varphi(z)|^{1/2}\,|dz|$, associated with a quadratic differential φ, has a natural generalization. Suppose f is holomorphic in a domain D of the z-plane and λ is a positive real number. We introduce the line element $|f(z)|^\lambda\,|dz|$. In order to make it a conformal metric, i.e. invariant under conformal mapping, the weight function f has to be transformed in a correct manner. Let $z = h(\tilde{z})$ be a conformal mapping of a domain $\tilde{D}$ into D. Let $\tilde{f}(\tilde{z})$ be such that

$$|\tilde{f}(\tilde{z})|^\lambda\,|d\tilde{z}| = |f(z)|^\lambda\,|dz|.$$

Then, $|\tilde{f}(\tilde{z})| = |f(z)| \cdot \left|\dfrac{dz}{d\tilde{z}}\right|^{1/\lambda}$, and hence necessarily $\tilde{f}(\tilde{z}) = f(z)\, h'(\tilde{z})^{1/\lambda}$. Conversely, if $\tilde{D}$ is simply connected, one can choose a single valued branch of $h'(\tilde{z})^{1/\lambda}$ and thus find $\tilde{f}(\tilde{z})$. In particular, $\tilde{f}$ can be determined locally in $\tilde{D}$.

Just as in the case of a quadratic differential one can introduce, locally, a distinguished or natural parameter, in terms of which the representation of the metric becomes particularly simple. In the neighborhood of a point z_0 with $f(z_0) \neq 0$, the integral $w = F(z) = \int_{z_0}^{z} f(t)^\lambda\, dt$ provides a conformal mapping with $|dw| = |f(z)|^\lambda\,|dz|$. In the w-plane, the length element is therefore the Euclidean element $|dw|$. We conclude, that any two points which are sufficiently close to z_0 can be joined by a unique shortest arc which satisfies $\arg dw = \arg f(z)^\lambda\, dz = \text{const}$. We call this a straight arc, and a maximal straight arc is called a trajectory. Of course, we can no longer speak of horizontal and vertical trajectories.

In the neighborhood of a zero z_0 of f of order n one can introduce a local parameter ζ such that the line element has the form $|dw| = |f(z)|^\lambda\,|dz| = |\zeta|^{n\lambda}\,|d\zeta|$. Any two points in a sufficiently small neighborhood of z_0 can again be joined by a unique shortest arc. If it passes through z_0, both of its angles must be $\geq \dfrac{\pi}{n\lambda + 1}$ (angle condition).

A polygon with respect to the metric $|f(z)|^\lambda\,|dz|$ is a curve composed of straight arcs. If it passes through zeroes of f, they are located at its vertices. A geodesic is a polygon which satisfies the angle condition at its vertices. Teichmüller's lemma for a simple closed polygon γ now reads

$$\sum_j \{\pi-(\lambda n_j+1)\vartheta_j\}=2\pi+2\pi\lambda\sum_i n_i\geq 2\pi,$$

where the $n_j\geq 0$ are the orders of the zeroes of f at the vertices of γ, ϑ_j the interior angles at these vertices, and n_i the orders of the enclosed zeroes. One concludes that there is no simple closed polygon γ which satisfies the angle condition at all but at most two of its vertices, if f is holomorphic in the interior of γ. It follows immediately that in a simply connected domain any two points can be joined by at most one geodesic. The same holds in an arbitrary domain with respect to the connecting arcs in a fixed homotopy class.

In a simply connected domain there can be no closed geodesic. Using the uniqueness of geodesic connections, one finds that every (maximal) geodesic ray must tend to the boundary.

The existence of a shortest connection is in general not assured. But in a relatively compact subregion the boundary of which is a simple polygon (with respect to $|f(z)|^\lambda|dz|$) there exists a unique relatively shortest connection of two arbitrary points. Again we conclude that in a simply connected domain every geodesic (i.e. locally shortest) arc is the unique shortest connection between any two of its points. The same is true in an arbitrary domain, if the connecting arcs are homotopic to the given geodesic arc.

The divergence principle holds as in the case of a quadratic differential: Let β be a geodesic arc with endpoints z_1 and z_2, and let α_1 and α_2 be geodesic rays with initial points z_1 and z_2 respectively. Suppose that the angles between the rays α_i and β are at least $\pi/2(\lambda n_i+1)$, where $n_i\geq 0$ is the order of the zero of f at z_i, $i=1,2$. Then the length of any curve γ joining α_1 to α_2 is $|\gamma|\geq|\beta|$.

The area element of the metric is $|f(z)|^{2\lambda}dx\,dy$. Let f be holomorphic in the unit disk D and let the total area

$$|D|=\iint_D |f(z)|^{2\lambda}dx\,dy$$

be finite. Then, every geodesic ray γ converges to a well determined boundary point $\zeta\in\partial D$. The same still holds under the less restrictive condition that there is a dense set of points $\zeta\in\partial D$ which can be joined to $z=0$ (or any interior point of D) by an arc of finite length in the metric $|f(z)|^\lambda|dz|$.

The proofs are practically the same as in the case of a quadratic differential ($\lambda=\frac{1}{2}$). For details see Strebel [14].

19.9 The isoperimetric inequality. T. Carleman [1] has given a very beautiful proof of the isoperimetric inequality $A\leq\dfrac{1}{4\pi}L^2$ on minimal surfaces, reducing it to a theorem on analytic functions in the unit disk. His argument can be applied to the metric $|f(z)|^\lambda|dz|$, $\lambda>0$. It consists of two parts:

The first part is a proof of the inequality by direct computation for the special metric $|g(z)|^2|dz|$. In the second part the general case is reduced to the above using Blaschke products. We state the first part as

Lemma 19.9. *Suppose $g \not\equiv 0$ is in the Hardy class H^2 of the unit disk D. Then, the isoperimetric inequality*

(1) $$A \leq \frac{1}{4\pi} L^2$$

holds for the metric $|g(z)|^2 |dz|$, with

$$A = \iint_D |g(z)|^4 \, dx \, dy$$

the area of D and

$$L = \int_{\partial D} |g(z)|^2 |dz|$$

the length of its circumference. Equality occurs for the functions of the form

(2) $$g(z) = \frac{c}{1 - qz}, \qquad |q| < 1,$$

only.

Proof. We set

$$g(z) = \sum_0^\infty a_n z^n, \quad g^2(z) = \sum_0^\infty b_n z^n, \quad b_n = a_0 a_n + a_1 a_{n-1} + \ldots + a_n a_0.$$

Then

(3) $$L = \int_0^{2\pi} |g(e^{i\vartheta})|^2 \, d\vartheta = \lim_{r \to 1} \int_0^{2\pi} |g(r \cdot e^{i\vartheta})|^2 \, d\vartheta$$
$$= \lim_{r \to 1} \int_0^{2\pi} \left(\sum_0^\infty a_n z^n \right) \left(\sum_0^\infty \bar{a}_n \bar{z}^n \right) d\vartheta$$
$$= \lim_{r \to 1} 2\pi \sum_0^\infty |a_n|^2 r^{2n} = 2\pi \sum_0^\infty |a_n|^2.$$

(4) $$A = \iint_D |g(z)|^4 r \, dr \, d\vartheta = \int_0^1 \int_0^{2\pi} \left(\sum_0^\infty b_n z^n \right) \left(\sum_0^\infty \bar{b}_n \bar{z}^n \right) r \, d\vartheta \, dr$$
$$= 2\pi \int_0^1 \sum_0^\infty |b_n|^2 r^{2n+1} \, dr = \pi \sum_0^\infty \frac{|b_n|^2}{n+1}.$$

The Schwarz inequality yields

(5) $$|b_n|^2 \leq (n+1)(|a_0|^2 |a_n|^2 + |a_1|^2 |a_{n-1}|^2 + \ldots + |a_n|^2 |a_0|^2).$$

Therefore

(6) $$A = \pi \sum_0^\infty \frac{|b_n|^2}{n+1} \leq \pi \sum_0^\infty (|a_0|^2 |a_n|^2 + \ldots + |a_n|^2 |a_0|^2)$$
$$= \pi \left(\sum_0^\infty |a_n|^2 \right)^2 = \frac{1}{4\pi} L^2,$$

which proves (1).

Let equality hold. Then, it must hold in (5) for all n, and hence

(7) $$a_0 a_n = a_1 a_{n-1} = \ldots = a_n a_1, \qquad n = 1, 2, 3, \ldots.$$

If $a_0=0$, $a_0 \cdot a_{2n}=a_n^2=0$ for every n, hence $g=0$. Setting $\frac{a_1}{a_0}=q$ if $a_0 \neq 0$, we find $\frac{a_n}{a_{n-1}}=q$, hence

$$a_n=a_0 \cdot q^n, \qquad n=1,2,3,\ldots, \tag{8}$$

proving (2) for $c=a_0$. Necessarily $|q|<1$, because of the convergence of the series $g(z)=a_0 \sum_0^\infty (qz)^n$ for $|z|<1$.

Conversely, if g is of the form (2), $g(z)=c \cdot \sum_0^\infty (qz)^n$, $g^2(z)=c^2 \sum_0^\infty (n+1)(qz)^n$. Therefore $a_n=c \cdot q^n$, $b_n=c^2(n+1)q^n$, $L=2\pi|c|^2 \sum_0^\infty |q|^{2n}$, $A=\pi|c|^4 \sum_0^\infty (n+1)|q|^{2n}$ $=\frac{1}{4\pi}L^2$. This proves the lemma.

We can now show the isoperimetric inequality for the metric $|f(z)|^\lambda |dz|$, $\lambda>0$, in the disk.

Theorem 19.9. *Let f be in the Hardy space H^λ, $0<\lambda<\infty$, of the disk. Then,* (1) *holds with*

$$A=\iint_D |f(z)|^{2\lambda}\,dx\,dy,$$

$$L=\int_{\partial D} |f(z)|^\lambda |dz|.$$

Equality occurs if and only if $f^{\frac{\lambda}{2}}$ is of the form (2).

Proof. By the F. Riesz factorization theorem (Duren [1]) the Blaschke product

$$B(z)=z^m \prod_n \frac{|z_n|}{z_n} \frac{z_n-z}{1-\bar{z}_n z},$$

formed with the zeroes 0 (of order m) and $z_n \neq 0$ of f converges, and $|B(e^{i\vartheta})|=1$ for a.a. values of $\vartheta \in [0,2\pi]$. Moreover, the quotient

$$\tilde{f}(z)=\frac{f(z)}{B(z)},$$

which has no zeroes, is in H^λ. We therefore have, with $g=\tilde{f}^{\frac{\lambda}{2}}$, $g \in H^2$, and

$$\begin{aligned} A &= \iint_D |f(z)|^{2\lambda}\,dx\,dy \le \iint_D |\tilde{f}(z)|^{2\lambda}\,dx\,dy \\ &= \iint_D |g(z)|^4\,dx\,dy \le \frac{1}{4\pi}\Big(\int_{\partial D} |g(z)|^2 |dz|\Big)^2 \\ &= \frac{1}{4\pi}\Big(\int_{\partial D} |\tilde{f}(z)|^\lambda |dz|\Big)^2 = \frac{1}{4\pi}\Big(\int_{\partial D} |f(z)|^\lambda |dz|\Big)^2 \\ &= \frac{1}{4\pi}L^2. \end{aligned}$$

In the case of equality, f can have no zeroes and g must be of the form (2). The converse is also true.

This proves the theorem. It shows, in particular, that the space of holomorphic f in D such that $|f|^{2\lambda}$ has finite area integral over D contains the Hardy space H^λ.

The isoperimetric inequality for arbitrary simply connected domains follows by conformal mapping.

Corollary 19.9. *Let f be holomorphic in the domain G of the z-plane, $\lambda>0$. Suppose G_0 is a relatively compact subdomain of G bounded by a rectifiable Jordan curve Γ. Then, the area*

$$A=\iint_{G_0}|f(z)|^{2\lambda}dx\,dy$$

of G_0 and the length

$$L=\int_{\Gamma}|f(z)|^{\lambda}|dz|$$

of Γ, measured in the metric $|f(z)|^{\lambda}|dz|$, satisfy the isoperimetric inequality (1). *Equality holds if and only if the function*

$$F(z)=\int f(z)^{\lambda}dz \tag{10}$$

is a $1-1$-conformal mapping of G_0 onto a disk.

Proof. Let $z=h(w)$ be a conformal mapping of the disk D: $|w|<1$, onto G_0. By the transformation rule we define

$$\tilde{f}(z)=f(z)\left(\frac{dz}{dw}\right)^{\frac{1}{\lambda}},$$

with any single valued branch of $h'(w))^{1/\lambda}$. Since $h'\in H^1(D)$ (see Duren [1]) and f is bounded on $\bar{G}_0$, $\tilde{f}\in H^{\lambda}(D)$. Applying Theorem 19.9 yields

$$\begin{aligned}\iint_{G_0}|f(z)|^{2\lambda}dx\,dy&=\iint_{D}|\tilde{f}(w)|^{2\lambda}du\,dv\\&\le\frac{1}{4\pi}\Big(\int_{\partial D}|\tilde{f}(w)|^{\lambda}|dw|\Big)^2\\&=\frac{1}{4\pi}\Big(\int_{\Gamma}|f(z)|^{\lambda}|dz|\Big)^2.\end{aligned}$$

Let equality hold. Then, $\tilde{f}^{\frac{\lambda}{2}}$ must be of the form (2), hence

$$F(w)=\int\tilde{f}(w)^{\lambda}dw=\frac{c}{1-qw}+d, \tag{12}$$

which is a linear transformation and therefore maps D onto a disk. But, because of the transformation rule,

$$\int f(z)^{\lambda}dz=\int\tilde{f}(w)^{\lambda}dw,$$

hence (10) maps G_0 onto a disk. Conversely, if this is true, (12) must hold and therefore $\tilde{f}(w)^{\frac{\lambda}{2}}$ is of the form (2) and we have equality in (1). This finishes the proof of the corollary.

Chapter VI. Quadratic Differentials with Closed Trajectories

§ 20. Extremal and Uniqueness Properties

Introduction. Quadratic differentials with closed trajectories were first considered by Teichmüller in his "Habilitationsschrift" [1] in the following example: two non overlapping punctured disks G' and G'', with punctures at $z=0$ and $z=\infty$ respectively, which do not contain $z=-1$, have to be chosen in such a way that their reduced moduli M' and M'' maximize the sum $q^2 M' + M''$. The solution ("Extremalgebiete des speziellen Modulsatzes", pg 33) is given by a quadratic differential with second order poles at 0 and ∞. For Riemann surfaces and finitely many punctures, the problem was solved by the author in [5] and [11], see § 23 below.

Jenkins [2] posed and solved a generalized extremal length problem on finite Riemann surfaces. He showed that this is equivalent with the problem of maximizing a weighted sum of moduli $\sum a_i^2 M_i$. He then used Schiffer's interior variation and Hadamard's variation to find that the solution is given by a quadratic differential with closed trajectories. The result is true for arbitrary Riemann surfaces (Jenkins and Suita [1], Strebel [11]; see Sections 21.10 and 21.13 below).

In connection with quasiconformal mappings, one would rather have moduli only and no lengths involved. This leads to the "moduli problem" which was solved by the author (Strebel [4] and [6]), using the methods of Jenkins and Schiffer. The solution of this problem which is presented in Sections 21.7 and 21.8 is based on the theorem about the heights of the cylinders. The generalization to open surfaces is in Strebel [11].

The "heights problem", which was posed and solved in full generality by Renelt [1], and, simultaneously but less general, by Hubbard and Masur [1] leads to a quadratic differential with closed trajectories for which the heights of the cylinders are prescribed. The solution is by quasiconformal deformation and Weyl's lemma. An improvement of Renelt's original method (Dirichlet Principle) was proposed by E. Reich (see D.R. Goodman [1]). (Sections 21.1 and 22.1 below.)

For an access to the problems from Teichmüller theory see F. Gardiner [1], [2].

§ 21 deals with the above mentioned problems for finite curve systems. The quintessence is, that they are all equivalent.

In § 22, some of the theorems are generalized to infinite curve systems.

20.1 Definition of a quadratic differential with closed trajectories. Quadratic differentials with closed trajectories can be defined on arbitrary Riemann surfaces and independently of their norm, although we shall mostly speak about quadratic differentials of finite norm.

Definition 20.1. A meromorphic quadratic differential $\varphi \neq 0$ on a Riemann surface R is said to have closed trajectories, if its non closed trajectories cover a set of measure zero. (A point set of R is said to have measure zero, if its intersection with every parameter neighborhood has areal measure zero in the respective parameter plane; this property is evidently independent of the choice of the parameter.)

A quadratic differential with closed trajectories cannot have poles of order higher than two, and the leading coefficient at every pole of order two must be negative.

Theorem 20.1. *Let R be a compact Riemann surface of genus $g \geq 2$ and let φ be a meromorphic quadratic differential of finite norm on R. Then φ has closed trajectories if and only if its critical graph Γ (set of critical trajectories together with their critical endpoints) is compact.*

Proof. Since φ has finite norm, all its critical points are either zeroes or poles of the first order.

Let Γ be compact. It cannot be empty since $g \neq 1$. The set of points of R of φ-distance $<\delta$ from Γ is, for sufficiently small positive δ, a neighborhood U of Γ the boundary of which consists of closed trajectories. Let now α be a trajectory of φ through a point $P \notin U$, and assume that it is not closed. Let α^+ be one of its rays. It is recurrent and therefore necessarily dense in a domain $D \subset R$. Since it cannot enter U, D is a proper subdomain of R bounded away from the critical points of φ, which is impossible, because the boundary of D necessarily contains critical points (see Corollary (2), Section 11.2). As we can choose $\delta > 0$ arbitrarily small, every regular trajectory of φ is closed.

Conversely, assume that φ has closed trajectories. Let α be a critical trajectory. If a ray α^+ of α does not tend to a critical point of φ, it is necessarily recurrent, since it cannot tend to the boundary nor to an infinite critical point of φ. But then, the interior of its closure is a domain which is disjoint from the closed trajectories of φ and has positive area.

20.2 Ring domains of a given homotopy type. A closed trajectory α of an arbitrary quadratic differential φ is embedded in a largest ring domain R_0, which is swept out by the closed trajectories of φ homotopic to α (Section 9.4). R_0 is uniquely determined by α except for a holomorphic φ on the torus, which is therefore excluded in the following. For a holomorphic φ, two closed trajectories α_1 and α_2 are freely homotopic if and only if the associated ring domains R_1 and R_2 coincide. The same is of course true for a meromorphic φ on the punctured surface $\dot{R} = R \setminus \{\text{poles of } \varphi\}$. If φ is a quadratic differential with closed trajectories, its characteristic ring domains (i.e. the maximal ring domains swept out by its closed trajectories) cover the surface R up to a set of measure zero. For a holomorphic quadratic differential on a compact surface of genus $g \geq 2$, the number of characteristic ring domains is at most $3g-3$.

A system of finitely or infinitely many Jordan curves γ_i on a Riemann surface R is called admissible, if none of the curves is homotopically trivial (homotope zero) and if, for $i \neq k$, the two curves γ_i and γ_k do not intersect and are not freely homotopic (Section 2.6).

If φ is holomorphic on R and if we pick a closed trajectory α_i from every characteristic ring domain R_i of φ, we have an admissible curve system $\{\alpha_i\}$. The same is true for a meromorphic φ if we puncture R at the poles of φ.

Definition 20.2. A ring domain R_0 on R is said to be of homotopy type γ, if a Jordan curve $\gamma_0 \subset R_0$ which separates its two boundary components is freely homotopic to γ.

A system of non overlapping ring domains $R_i \subset R$ is said to be of homotopy type $\{\gamma_i\}$, where $\{\gamma_i\}$ is an admissible curve system on R, if every R_i is of homotopy type γ_i for exactly one γ_i. It is convenient to allow for "degenerate" ring domains, i.e. non existing ones. We attribute them the modulus zero.

Finally, a holomorphic quadratic differential φ with closed trajectories is said to be of homotopy type $\{\gamma_i\}$, if its characteristic ring domains are of this type. In the case of a meromorphic differential φ with closed trajectories, the curves γ_i have to lie on the surface punctured at the poles of φ and homotopy is meant with respect to the punctured surface.

20.3 Extremality of the φ-metric. Quadratic differentials with closed trajectories have several extremal properties. They are all based on integrating the length inequality for closed curves over annular regions and applying the Schwarz inequality.

The characteristic ring domains R_i of a quadratic differential φ with closed trajectories cover the whole surface R, up to a set of measure zero. This allows for a simple representation of the norm of φ. A domain R_i, cut along a vertical arc connecting its two boundary components, is mapped by a branch $w=\Phi(z)$ of $\int \sqrt{\varphi}\, dz$ onto a horizontal rectangle S_i in the $w=u+iv$-plane. S_i has horizontal and vertical sides of length

(1) $$a_i = \int_{\alpha_i} |\varphi(z)|^{\frac{1}{2}}\, |dz|$$

and $b_i = a_i M_i$ respectively, with M_i the modulus of R_i. The horizontals in S_i correspond to the trajectories α_i, of φ-length a_i, sweeping out R_i. Since the representation of φ in terms of the natural parameter $w=u+iv$ is $\varphi(w)\equiv 1$, the norm of the restriction of φ to R_i is the Euclidean area of S_i:

(2) $$\|\varphi\|_{R_i} = \iint_{R_i} |\varphi(z)|\, dx\, dy = \iint_{S_i} du\, dv = a_i b_i.$$

A quadratic differential φ with finite norm has closed trajectories if and only if $\|\varphi\| = \sum a_i b_i$, where the sum is taken over all the maximal ring domains swept out by horizontal trajectories (resp. one such domain in the case of a torus).

The norm of φ can be expressed in three ways:

(3) $$\|\varphi\| = \sum a_i b_i = \sum a_i^2 M_i = \sum \frac{b_i^2}{M_i}.$$

We now consider a non negative, locally square integrable metric ρ. This means that its representation in terms of a local parameter z transforms such that the differential $\rho(z)|dz|$ is an invariant expression. Moreover, $\rho(z)$ is measurable and $\iint \rho^2(z)\,dx\,dy$, which is meaningful because of the transformation rule for ρ, is locally finite.

If ρ is a majorant of the φ-metric in the sense that $\int_{\alpha_i} \rho(z)\,dz \geq a_i$ for a.a. closed trajectories α_i and for all i, the length area method, applied to the characteristic ring domains R_i of φ, immediately leads to the inequality

(4) $$\|\varphi\| \leq \iint_R \rho^2\,dx\,dy,$$

with equality iff $\rho = |\varphi|^{\frac{1}{2}}$ a.e. However, it is important to apply the method to arbitrary non overlapping ring domains $\tilde{R}_i$ of a given homotopy type and not just the characteristic ring domains of a quadratic differential.

To this end, let (γ_i) be an admissible system of Jordan curves on R. Assign numbers $a_i > 0$ to the free homotopy classes determined by the simple loops γ_i. A locally square integrable metric ρ is said to be admissible for the curve system (γ_i) with lengths a_i, if

(5) $$a_i \leq \int_\gamma \rho(z)\,|dz|$$

for a.a. γ freely homotopic to γ_i, in the following sense: For every ring domain $\tilde{R}_i$ of R of homotopy type γ_i inequality (5) holds for a.a. concentric circles of a circular annulus which is conformally equivalent to $\tilde{R}_i$.

Theorem 20.3 (Jenkins [2], [3]). *Given an admissible system of Jordan curves γ_i on an arbitrary Riemann surface R, and assigned lengths $a_i > 0$. Let ρ be a metric admissible for the curves γ_i and lengths a_i. Suppose, $\tilde{R}_i$ is a system of non overlapping ring domains of homotopy type (γ_i), with moduli $\tilde{M}_i \geq 0$. Then,*

(6) $$\sum_i a_i^2 \tilde{M}_i \leq \iint_R \rho^2\,dx\,dy,$$

and hence

(6′) $$\sup_{\{(\tilde{R})\}} \sum_i a_i^2 \tilde{M}_i \leq \inf_{\{\rho\}} \iint_R \rho^2\,dx\,dy.$$

Proof. We assume that the right hand side of (6) is finite. Let $\tilde{R}_i$ be a non degenerate ring domain (we forget about the degenerate ones). We map it conformally onto a circular annulus. The latter, cut along a radius, is then mapped onto a horizontal rectangle $\tilde{S}_i$, the concentric circles of the annulus corresponding to the horizontals of $\tilde{S}_i$. The mapping is normalized by the requirement that the horizontal side of $\tilde{S}_i$ has length a_i. Its vertical side

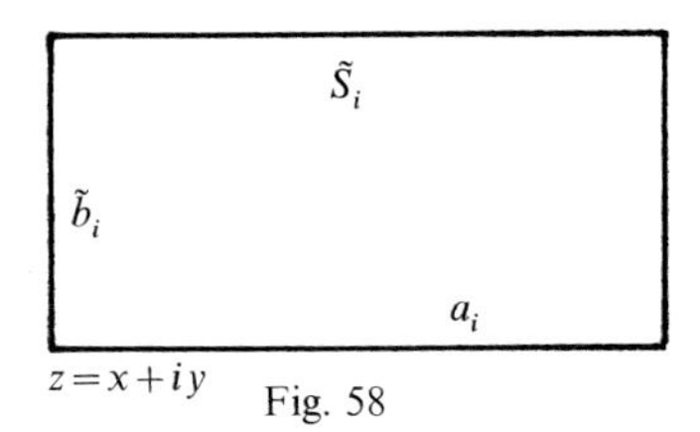

Fig. 58

becomes $\tilde{b}_i = a_i \tilde{M}_i$ (Fig. 58). We now have

(7) $$a_i \leq \int \rho(x+iy)\,dx$$

for a.a. values of y in $\tilde{S}_i$. Integration over y and summation over i yield

(8) $$\sum_i a_i \tilde{b}_i \leq \sum_i \iint_{S_i} \rho(x+iy)\,dx\,dy = \iint_{\cup S_i} \rho(z)\,dx\,dy.$$

Note that the last integral is not invariant under a change of the parameter, but we can arrange the $\tilde{S}_i$ disjointly in the z-plane and thus get, in fact, an integral over a subset of the z-plane. We apply the Schwarz inequality to (8) and thus get

(9) $$\Big(\sum_i a_i \tilde{b}_i\Big)^2 \leq \sum_i a_i \tilde{b}_i \cdot \iint_{\cup S_i} \rho(z)^2\,dx\,dy,$$

and hence

(10) $$\sum a_i \tilde{b}_i \leq \iint_{\cup \tilde{R}_i} \rho(z)^2\,dx\,dy \leq \iint_R \rho^2(z)\,dx\,dy,$$

which proves (6).

The extremality of the φ-metric now follows immediately by specialization. We start with a quadratic differential with closed trajectories and apply (6) to its characteristic ring domains R_i. We set $a_i = |\alpha_i|_\varphi$ and remember that the conformal mappings $R_i \to S_i$ are equal to Φ. We have $\tilde{b}_i = b_i$, the height of the cylinder R_i.

Corollary 20.3. *Let φ be a quadratic differential with closed trajectories on R. Assume that the metric $\rho|dz|$ is a majorant of the metric $|\varphi|^{\frac{1}{2}}|dz|$ in the sense that $\int_{\alpha_i} \rho(z)\,|dz| \geq a_i$ for a.a. α_i and all i. Then*

(11) $$\sum_i a_i b_i = \|\varphi\| \leq \iint_R \rho^2\,dx\,dy.$$

If the right hand side is finite, equality holds iff $\rho = |\varphi|^{\frac{1}{2}}$ a.e.

The proof follows immediately from (6), using (3). The statement about the equality sign is a consequence of the use of Schwarz's inequality and the fact that $\varphi \equiv 1$ in terms of the natural parameter.

The metric $|\varphi|^{\frac{1}{2}}|dz|$ is therefore uniquely extremal in the following sense: For all majoring metrics $\rho|dz|$ we have $\|\varphi\| \leq \iint_R \rho^2\,dx\,dy$ with equality iff the metrics are the same a.e. (The uniqueness of the extremal metric also follows, in the most general sense, by a convexity argument.)

20.4 Weighted sum of moduli. We can apply (6) to the metric induced by a quadratic differential φ of finite norm instead of ρ. This is particularly interesting if φ has closed trajectories, but in the first part of the following theorem φ is arbitrary.

Theorem 20.4. *Let $\varphi = 0$ be a holomorphic quadratic differential of finite norm on an arbitrary Riemann surface R. Suppose (γ_i) $i=1,2,3,\ldots$ is an admissible system of Jordan curves on R with the property that the ring domains of homotopy type γ_i have moduli $\leq \bar{M}_i < \infty$ for all i. Let*

$$a_i = \inf_{\gamma \sim \gamma_i} \int_\gamma |\varphi(z)|^{\frac{1}{2}} |dz|, \tag{1}$$

where γ varies over the rectifiable loops in the free homotopy class of γ_i. Then, for any system of non overlapping ring domains $\tilde{R}_i$ of homotopy type (γ_i) we have

$$\sum a_i^2 \tilde{M}_i \leq \|\varphi\|. \tag{2}$$

Equality holds iff the ring domains $\tilde{R}_i$ cover R up to a set of measure zero and are swept out by parallel θ-trajectories of φ, for some fixed θ.

If, in particular, φ has closed trajectories of homotopy type (γ_i), with characteristic ring domains R_i of moduli M_i, then

$$\sum a_i^2 \tilde{M}_i \leq \sum a_i^2 M_i, \tag{3}$$

with equality iff $\tilde{R}_i = R_i$ for all i, $\theta = 0$.

Proof. It follows from $\bar{M}_i < \infty$ that $a_i > 0$ for all i. We can therefore apply Theorem 20.3 and thus get

$$\sum a_i^2 \tilde{M}_i \leq \|\varphi\|. \tag{4}$$

Assume that the equality sign holds. Then, by the Schwarz inequality, $|\varphi(z)| = \text{const}$, and the constant must be equal to one. Therefore $\varphi_k(z) = e^{i\theta_k}$, where $\varphi_k(z)$ is the representation of φ in terms of the parameter z in $\tilde{S}_k$. Moreover, $R \setminus \bigcup \tilde{R}_k$ must have zero area.

It is almost evident, that ϑ_k cannot depend on k. To prove it, fix a ring, $\tilde{R}_1$ say. Pick a point $P_1 \in \tilde{R}_1$ and connect P_1 with a point $P_i \in \tilde{R}_i$ of an arbitrary ring domain $\tilde{R}_i$ by a geodesic polygon γ. We can choose γ such that it does not pass throught a zero of φ, that all its finitely many vertices lie in ring domains $\tilde{R}_k$ and that every side of it is contained in a φ-disk. Consider the first side of γ. Let its endpoints be $P_1 \in \tilde{R}_1$ and $P_2 \in \tilde{R}_2 \neq \tilde{R}_1$. Let α_1 and α_2 be the distinguished geodesics through P_1 and P_2 in $\tilde{R}_1$ and $\tilde{R}_2$ respectively (Fig. 59). By

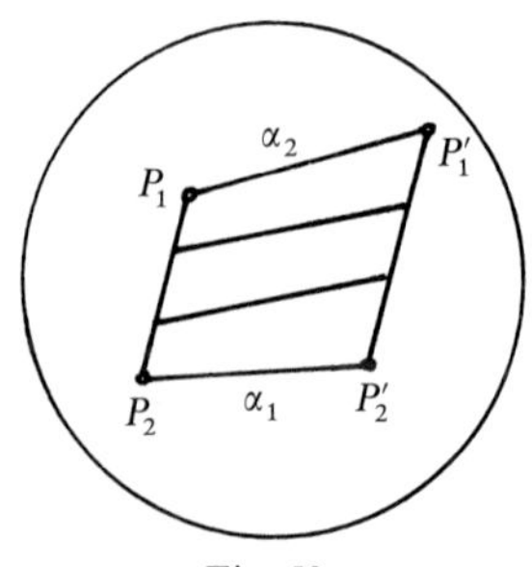

Fig. 59

assumption they have directions $\frac{1}{2}\vartheta_k$, $k=1,2$. Choose P_1' and P_2' on α_1 and α_2 respectively such that the two points can be joined by a straight segment parallel to the arc P_1P_2. Every ring domain $\tilde{R}_j$ which meets P_1P_2 also meets $P_1'P_2'$ and its intersection with the quadrangle with vertices P_1, P_2, P_2', P_1' is a parallelogram. The sum of their sides on P_1P_2 is equal to the φ-length $|P_1P_2|$, and similarly for the sides on $P_1'P_2'$. Therefore $|P_1P_2|=|P_1'P_2'|$, and we conclude that P_1P_1' is parallel to P_2P_2'. Continuing along γ we see that $\vartheta_i=\vartheta_1$. This proves the first part of the theorem. Of course the ring domains $\tilde{R}_i$ are uniquely determined (except for the torus).

Let now φ have closed horizontal trajectories of homotopy type (γ_i), with characteristic ring domains R_i. Then $\|\varphi\|=\sum a_i b_i=\sum a_i^2 M_i$, thus proving (**). Let equality hold. Assume first that for some γ_i both domains R_i and $\tilde{R}_i$ are not degenerate. Then a_i is the length of the closed trajectories in R_i and, since we must have equality in (1), $\vartheta=0$. This shows that the $\tilde{R}_i$ are in fact the characteristic ring domains of φ, and hence the two systems are the same: $\tilde{R}_i=R_i$, for all i.

If there is no such γ_i with both R_i and $\tilde{R}_i$ non degenerate, we pick R_i and $\tilde{R}_k$ such that $R_i\cap\tilde{R}_k\neq\emptyset$. Note that both sets (R_i) and $(\tilde{R}_i)$ cover R up to a set of measure zero. Then $\gamma_i\nsim\gamma_k$, and also $\gamma_i\cap\gamma_k=\emptyset$. Let α be a closed trajectory in R_i and let $\tilde{\alpha}$ be a closed geodesic in $\tilde{R}_k$. The two curves necessarily intersect, since they cannot be parallel; but their intersection number is zero. We can therefore find two segments on α and $\tilde{\alpha}$ respectively which bound a simply connected domain on R. That contradicts Teichmüller's lemma. The only way out is that the former case holds. That proves the theorem.

20.5 Weighted sum of the reciprocals of moduli. The norm of a quadratic differential with closed trajectories can be written in the form

$$\|\varphi\|=\sum a_i b_i=\sum\frac{b_i^2}{M_i},$$

where the coefficients b_i are the heights of the cylinders R_i in the φ-metric. This leads to the following inequality.

Theorem 20.5. *Let $\varphi\neq 0$ be a holomorphic differential with closed trajectories and finite norm on an arbitrary Riemann surface R. (The meromorphic case is again reduced to this one by introducing the punctured surface $\dot{R}=R\{$poles of $\varphi\}$ instead of R.) Let $\tilde{R}_i$ be a system of non overlapping ring domains on R with the homotopy type of the system of characteristic ring domains $\{R_i\}$ of φ. Then, their moduli $\tilde{M}_i$ satisfy*

$$\sum\frac{b_i^2}{\tilde{M}_i}\geq\sum\frac{b_i^2}{M_i},$$

with equality iff $\tilde{R}_i=R_i$ for every i.

Proof. If there is a degenerate ring domain $\tilde{R}_i$ ($\tilde{M}_i=0$), there is nothing to prove, since $b_i>0$ for every i. We therefore assume that $\tilde{M}_i>0$ for all i. As in the preceeding section we map $\tilde{R}_i$, after cutting it radially, onto a horizontal

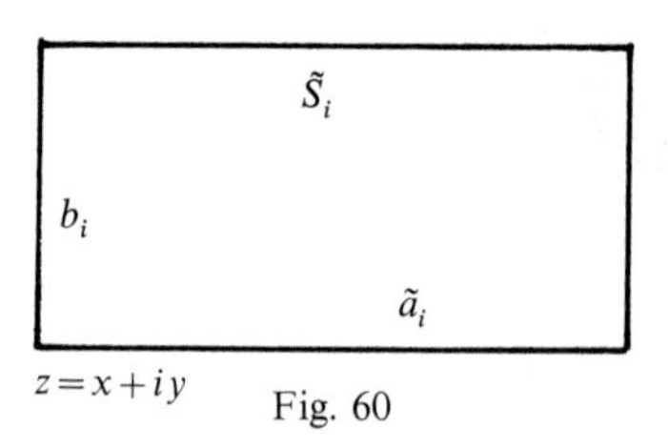

Fig. 60

rectangle $\tilde{S}_i$ of the $z=x+iy$-plane (Fig. 60). This time, $\tilde{S}_i$ is normalized to have height b_i and length $\tilde{a}_i=\frac{b_i}{\tilde{M}_i}$. Using z as parameter (and representing φ in terms of z) we get as before

$$a_i \leq \int |\varphi(x+iy)|^{\frac{1}{2}}\, dx, \tag{1}$$

since every horizontal of $\tilde{S}_i$ corresponds to a closed curve in the free homotopy class of α_i. We get by integration, summation and application of the Schwarz inequality

$$a_i b_i \leq \iint_{\tilde{S}_i} |\varphi(z)|^{\frac{1}{2}}\, dx\, dy \tag{2}$$

$$\sum a_i b_i \leq \sum \iint_{\tilde{S}_i} |\varphi(z)|^{\frac{1}{2}}\, dx\, dy = \iint_{\cup \tilde{S}_i} |\varphi(z)|^{\frac{1}{2}}\, dx\, dy \tag{3}$$

(note that the right hand integral, as previously, is not invariant under a change of the parameter),

$$\begin{aligned} (\textstyle\sum a_i b_i)^2 &\leq \textstyle\sum \tilde{a}_i b_i \cdot \displaystyle\iint_{\cup \tilde{S}_i} |\varphi(z)|\, dx\, dy \\ &\leq \textstyle\sum \tilde{a}_i b_i \cdot \displaystyle\iint_R |\varphi|\, dx\, dy = \textstyle\sum \tilde{a}_i b_i \cdot \sum a_i b_i. \end{aligned} \tag{4}$$

This proves the inequality

$$\sum a_i b_i = \sum \frac{b_i^2}{M_i} \leq \sum \frac{b_i^2}{\tilde{M}_i}. \tag{5}$$

Let equality hold. We conclude again that it must hold in (1) and therefore $\tilde{R}_i$ is a subannulus of R_i, swept out by closed trajectories. From (4) it follows that $\bigcup \tilde{S}_i = R$, and therefore $\tilde{R}_i = R_i$ for every i.

20.6 A minimax property of moduli. The following theorem has to do with moduli only, without any lengths involved. It is therefore meaningful for any quadratic differential with closed trajectories, whether of finite or of infinite norm. But it is proved here only for quadratic differentials of finite norm.

Theorem 20.6. *Let φ be a holomorphic quadratic differential with closed trajectories and finite norm on an arbitrary Riemann surface R. (The meromorphic case is reduced to this one by puncturing R as in the previous theorems.) Let $\{\tilde{R}_i\}$ be a system of ring domains on R of the same homotopy type as the system*

$\{R_i\}$ of characteristic ring domains of φ. Then their moduli $\tilde{M}_i$, M_i satisfy

$$\inf\left\{\frac{\tilde{M}_i}{M_i}\right\}\leq 1,$$

with equality iff $\tilde{R}_i = R_i$ for all i.

Proof. Let $q=\inf\left\{\frac{\tilde{M}_i}{M_i}\right\}$. We have

$$\tilde{M}_i \geq q\cdot M_i \tag{1}$$

for all i. Therefore, multiplication with a_i^2 and summation yields, by Theorem 20.4,

$$\sum a_i^2\,\tilde{M}_i \geq q\cdot \sum a_i^2\,M_i \geq q\cdot \sum a_i^2\,\tilde{M}_i. \tag{2}$$

If $q=0$, there is nothing to prove; otherwise $\sum a_i^2\,\tilde{M}_i>0$ and we get $1\geq q$.

Let equality hold. Then, (2) involves that $\sum a_i^2\,\tilde{M}_i=\sum a_i^2\,M_i$, and by Theorem 20.4 again we have $\tilde{R}_i=R_i$ for all i.

§21. Existence Theorems for Finite Curve Systems

There are three types of existence theorems for quadratic differentials with closed trajectories according to the data which are prescribed: One can give the lengths of the circumferences of the cylinders or their heights or the ratio of their moduli. The first problem is an extremal metric problem. Here, some of the ring domains can collaps, when the lengths are relatively small. This problem was solved by J.A. Jenkins [2]. The third problem was solved by the author [4], [6]. In this problem there is of course no degeneration of the ring domains possible, but the functional which is maximized is less easy to handle. The second problem seems to be the most adequate, in the sense that it combines both advantages: There is no degeneration and we have a simple functional to minimize. It was solved simultaneously by J. Hubbard and H. Masur [1] and by H. Renelt [1]. We first solve the second problem following Renelt's proof with a modification due to E. Reich (see Goodman [1]).

In this § we only deal with finite admissible curve systems and with quadratic differentials of finite norm. This is, in the case of finite curve systems, equivalent with the requirement that the ring domains of corresponding homotopy type have bounded moduli.

21.1 Existence for given heights of the cylinders

Theorem 21.1. *Let $(\gamma_1,\ldots,\gamma_p)$ be a finite admissible curve system on a Riemann surface R, which has the property that the ring domains of the given homotopy types γ_i have bounded moduli for all i. Then, for arbitrary positive numbers b_i, there exists a holomorphic quadratic differential φ on R with closed* *trajectories, the ring domains R_i of which have homotopy type γ_i and heights b_i, $i=1, \ldots, p$.*

φ is uniquely determined and has finite norm $\|\varphi\| = \sum b_i^2/M_i$, where M_i is the modulus of R_i.

Proof. For given positive numbers b_i, let $(R_{in})_{i=1,\ldots,p}$ be systems of non overlapping domains on R of homotopy type $(\gamma_i)_{i=1,\ldots,p}$, with moduli M_{in} and such that the sum

$$\sum_{i=1}^{p} \frac{b_i^2}{M_{in}}$$

tends to its infimum. Let g_{in} be a conformal mapping of the annulus $r_{in}<|z|<1$, $M_{in}=\frac{1}{2\pi}\log\frac{1}{r_{in}}$, onto R_{in}. It is clear that M_{in} cannot tend to zero, because otherwise $\frac{b_i^2}{M_{in}} \to \infty$. Since the supremum $\bar{M}_i$ of the moduli of all the ring domains of homotopy type γ_i is finite, the sequence $(g_{in})_{n=1,2,\ldots}$ is normal. We can therefore find a subsequence which converges locally uniformly in $r_i<|z|<1$, $r_i=\lim_{n\to\infty} r_{in}$, to a conformal mapping g_i of the annulus $r_i<|z|<1$ onto a ring domain R_i. R_i has modulus $M_i=\frac{1}{2\pi}\log\frac{1}{r_i}$ and homotopy type γ_i, and the system (R_i) is extremal in the sense that the corresponding sum

$$\sum_{i=1}^{p} \frac{b_i^2}{M_i}$$

is minimal.

We now perform a quasiconformal selfmapping of the surface R which is homotopic to the identity. It deforms the ring domains R_i into ring domains $\tilde{R}_i$ with moduli $\tilde{M}_i$ and a sum

$$\sum \frac{b_i^2}{\tilde{M}_i} \geq \sum \frac{b_i^2}{M_i}. \tag{1}$$

For an arbitrary quasiconformal mapping of R_i onto $\tilde{R}_i$ we estimate the change of the modulus. To this end we map both ring domains conformally onto circular annuli R_i' and $\tilde{R}_i'$ respectively (Fig. 61). Cutting R_i' radially and applying a suitable multiple of the logarithm we end up with a horizontal rectangle S_i of height b_i and length $a_i=b_i/M_i$.

The annulus $\tilde{R}_i'$ is cut along the image of the radial cut of R_i'. We then apply a similar procedure and end up with a horizontal quadrilateral of height b_i and length (horizontal distance of corresponding boundary points) $\tilde{a}_i$ (Figs. 61 and 62). The modulus of the circular ring $\tilde{R}_i'$ and hence of the original ring domain $\tilde{R}_i$ is $\tilde{M}_i=b_i/\tilde{a}_i$.

Let $\zeta^*=f(\zeta)$, $\zeta=\xi+i\eta$, be the induced quasiconformal mapping of S_i onto $\tilde{S}_i$. The length of the image $f(\alpha)$ of any horizontal α in S_i is at least $\tilde{a}_i$:

$$\tilde{a}_i \leq \int_{f(\alpha)} |d\zeta^*| = \int_{\alpha} |f_\zeta + f_{\bar{\zeta}}|\, d\xi. \tag{2}$$

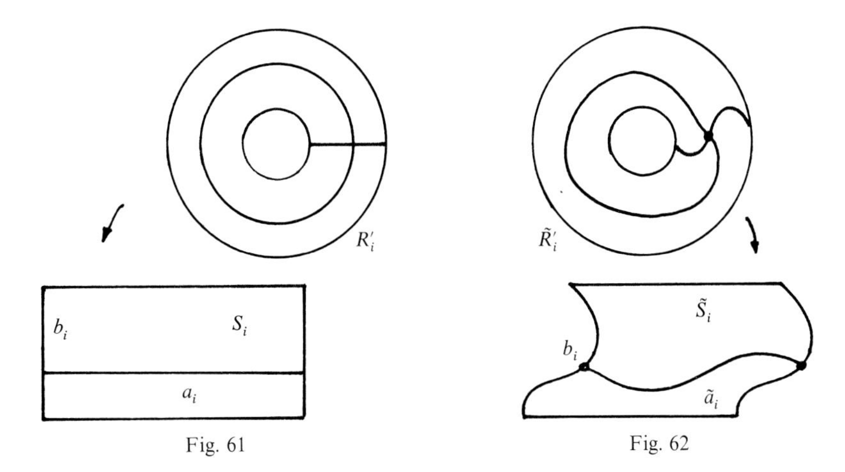

Fig. 61 Fig. 62

Integration over η and application of the Schwarz inequality yield successively

$$\tilde{a}_i b_i \le \iint_{S_i} |f_\zeta + f_{\bar{\zeta}}| \, d\xi \, d\eta = \iint_{S_i} J^{\frac{1}{2}} \frac{|f_\zeta + f_{\bar{\zeta}}|}{J^{\frac{1}{2}}} \, d\xi \, d\eta \tag{3}$$

with $J = |f_\zeta|^2 - |f_{\bar{\zeta}}|^2$ the Jacobian of f,

$$(\tilde{a}_i b_i)^2 \le \iint_{S_i} J \, d\xi \, d\eta \cdot \iint_{S_i} \frac{|f_\zeta + f_{\bar{\zeta}}|^2}{J} \, d\xi \, d\eta \tag{4}$$

$$= \tilde{a}_i b_i \cdot \iint_{S_i} \frac{|1+\kappa|^2}{1-|\kappa|^2} \, d\xi \, d\eta,$$

with $\kappa = f_{\bar{\zeta}}/f_\zeta$ the complex dilatation of f. We finally get

$$\frac{b_i^2}{\tilde{M}_i} = \tilde{a}_i b_i \le \iint_{S_i} \frac{|1+\kappa|^2}{1-|\kappa|^2} \, d\xi \, d\eta. \tag{5}$$

We can transform the right hand side into an integral over the original ring domain R_i and sum up to get an integral over the Riemann surface R. Let Φ_i be the conformal mapping of R_i (slit along the curve which corresponds to the radial slit in R_i') onto S_i. We put $\varphi(z) = \Phi_i'(z)^2$ in terms of a local parameter z in R_i, and $\varphi = 0$ in the complement $R \setminus \bigcup R_i$. We then have

$$\frac{b_i^2}{\tilde{M}_i} \le \iint_{R_i} \frac{|1+\kappa|^2}{1-|\kappa|^2} |\Phi_i'|^2 \, dx \, dy = \iint_{R_i} \frac{|1+\kappa|^2}{1-|\kappa|^2} |\varphi| \, dx \, dy \tag{6}$$

and

$$\sum \frac{b_i^2}{\tilde{M}_i} \le \sum \iint_{R_i} \frac{|1+\kappa|^2}{1-|\kappa|^2} |\varphi| \, dx \, dy = \iint_R \frac{|1+\kappa|^2}{1-|\kappa|^2} |\varphi| \, dx \, dy. \tag{7}$$

Of course, φ transforms like a quadratic differential under a change of the parameter, which makes the integral over R well defined.

We now specify the quasiconformal selfmapping of R. Let U be a parameter neighborhood on R with a parameter z. Choose an arbitrary complex valued function h with continuous first order derivatives and with compact support in U. The function $z \to w = z + \varepsilon h(z)$, continued by the identity outside of U, is a quasiconformal selfmapping of R, for all sufficiently small complex numbers ε. It is clearly homotopic to the identity and has the differential

(8) $$dw=(1+\varepsilon h_z)\,dz+\varepsilon h_{\bar z}\,d\bar z.$$

In R_i the conformal mapping Φ_i: $R_i \to S_i$ has the differential $d\zeta=\Phi_i'(z)\,dz$. In order to compute the complex dilatation κ of the induced quasiconformal mapping f_i: $S_i \to \tilde S_i$ we need the differential

(9) $$dw=(1+\varepsilon h_z)\frac{d\zeta}{\Phi_i'}+\varepsilon h_{\bar z}\frac{d\bar\zeta}{\overline{\Phi_i'}};$$

the conformal mapping of $\tilde R_i$ onto $\tilde S_i$ has no influence on the complex dilatation. We get

(10) $$\kappa=\frac{\varepsilon h_{\bar z}}{1+\varepsilon h_z}\frac{\Phi_i'}{\overline{\Phi_i'}}=\frac{\varepsilon h_{\bar z}}{1+\varepsilon h_z}\frac{\varphi}{|\varphi|}.$$

Simple computations show that $|\kappa|=O(\varepsilon)$,

(11) $$\begin{aligned}|\varphi|\frac{|1+\kappa|^2}{1-|\kappa|^2}&=|\varphi|\,(1+2\operatorname{Re}\kappa+O(|\kappa|^2))\\&=|\varphi|+2\operatorname{Re}\varepsilon h_{\bar z}\varphi+|\varphi|\cdot O(\varepsilon^2).\end{aligned}$$

The integral on the right hand side of (7) can be written in the form (since $\kappa=0$ outside of U)

(12) $$\begin{aligned}\iint_R \frac{|1+\kappa|^2}{1-|\kappa|^2}|\varphi|\,dx\,dy&=\iint_R|\varphi|\,dx\,dy+\iint_U\left(\frac{|1+\kappa|^2}{1-|\kappa|^2}-1\right)|\varphi|\,dx\,dy\\&=\sum a_i b_i+\iint_U\left(\frac{|1+\kappa|^2}{1-|\kappa|^2}-1\right)|\varphi|\,dx\,dy\\&=\sum\frac{b_i^2}{M_i}+2\operatorname{Re}\varepsilon\iint_U h_{\bar z}\varphi\,dx\,dy+O(\varepsilon^2).\end{aligned}$$

From (1) we conclude that $\iint_U h_{\bar z}\varphi\,dx\,dy=0$ for every continuously differentiable h with compact support in U. Weyl's lemma (Ahlfors [2], p. 45) shows that φ is a.e. equal to an analytic function in U. Proceeding in the same way along overlapping neighborhoods we realize that φ is a.e. equal to a holomorphic quadratic differential on R, which we may call φ again. We conclude, that $R\setminus\bigcup R_i$ is a null set. In R_i, $\varphi=\Phi_i'^2$. Therefore R_i is swept out by closed trajectories of φ. As R is, by assumption, different from the compact torus, the characteristic ring domains of φ are well determined. Every R_i clearly is a subannulus of such a characteristic ring domain (since it is swept out by closed trajectories). If one at least of the R_i were a proper subdomain of the associated characteristic ring domain, this would contradict the fact that $R\setminus\bigcup R_i$ is a null set. Therefore the R_i are the characteristic ring domains of φ, and their heights are the same as the heights of the rectangles S_i, i.e. equal to b_i. This finishes the proof of the existence theorem. The uniqueness was shown in Section 20.5.

In particular, if $p=1$, we have only one cylinder, of homotopy class γ_1. The corresponding quadratic differentials are called simple. They are determined by γ_1 up to positive multiples. The modulus $M_1 = \bar{M}_1$ of the cylinder is uniquely determined.

21.2 A compactness property. Let $(\gamma_1, \ldots, \gamma_p)$ be an admissible system of Jordan curves on an arbitrary Riemann surface R, such that the maximal modulus $\bar{M}_i$ is finite for every $i=1, \ldots, p$. The set of holomorphic quadratic differentials on R with closed trajectories and with ring domains R_i of homotopy type γ_i, $i=1, \ldots, p$ is denoted by Γ. It is admitted that some of the R_i are degenerate, and we define $\varphi = 0 \in \Gamma$.

Every $\varphi \in \Gamma$ has finite norm $\|\varphi\| = \sum_{i=1}^{p} a_i^2 M_i$. The subset of normalized elements of Γ with norm one is denoted by Γ_0.

Theorem 21.2. *The set Γ_0 of normalized quadratic differentials with closed trajectories of homotopy type $(\gamma_1, \ldots, \gamma_p)$ is compact.*

Proof. For a given admissible system of Jordan curves γ_i, $i=1, \ldots, p$, on R, let $\varphi_n \in \Gamma_0$, $n=1, 2, 3, \ldots$. By Cauchy's integral, the sequence $\varphi_n(z)$ is locally bounded for any fixed local parameter z. Therefore there exists a subsequence which converges locally uniformly to a holomorphic quadratic differential φ; we call this subsequence (φ_n) again. Because of the uniform convergence on compact sets, $\|\varphi\| \leq 1$.

Let $a_{in} = \inf_{\gamma \sim \gamma_i} \int_\gamma |\varphi_n(z)|^{\frac{1}{2}} |dz|$; if the ring domain R_{in} of φ_n is not degenerate, this is the length of the trajectories of φ_n which are homotopic to γ_i. For any fixed curve γ which is freely homotopic to γ_i we have

$$a_{in} \leq \int_\gamma |\varphi_n(z)|^{\frac{1}{2}} |dz| \to \int_\gamma |\varphi(z)|^{\frac{1}{2}} |dz|,$$

hence

$$\overline{\lim_{n \to \infty}}\, a_{in} \leq a_i = \inf_{\gamma \sim \gamma_i} \int_\gamma |\varphi(z)|^{\frac{1}{2}} |dz|.$$

Since $\|\varphi_n\| = \sum_{i=1}^{p} a_{in}^2 M_{in} = 1$, not all the moduli M_{in} can tend to zero for $n \to \infty$. Because $M_{in} \leq \bar{M}_i < \infty$ for all i and n, we can pass to another subsequence such that the sequences (M_{in}) converge. Let, by an appropriate choice of the numbering, $\lim_{n \to \infty} M_{in} = M_i > 0$ for $i=1, \ldots, q$ and $\lim_{n \to \infty} M_{in} = 0$ for $i=q+1, \ldots, p$.

The function $f_{in} = e^{\frac{2\pi i}{a_{in}} \Phi_n}$, $\Phi_n = \int \sqrt{\varphi_n}$, with a proper choice of the additive constant, is a conformal mapping of the ring domain R_{in} onto the annulus R'_{in}: $r_{in} < |z| < 1$, $M_{in} = \frac{1}{2\pi} \log \frac{1}{r_{in}}$. The inverse $g_{in} = f_{in}^{-1}$ maps R'_{in} onto R_{in}. The sequences of mappings $(g_{in})_{n=1,2,\ldots}$ form a normal family. We can therefore assume, by passing to another subsequence, that they converge locally uniformly in R'_i: $r_i < |z| < 1$, with $r_i = \lim_{n \to \infty} r_{in}$, to a conformal mapping g_i of R'_i onto a ring domain $R_i \subset R$.

In R_i we can introduce the parameter z by means of the inverse mapping $f_i = g_i^{-1}$, and we may identify the points $P \in R_i$ of R with their corresponding points $z \in R'_i$. Let now D_0 be any closed disk in R'_i. We choose a slightly larger concentric closed disk $D \subset R'_i$. For all sufficiently large n the disk D is contained in R'_{in}, and its image by g_{in} is contained in R_i. We can therefore, after the identification, consider g_{in} as a conformal mapping of D into R'_i, and the sequence (g_{in}) converges to the identity uniformly on D. For every $\varepsilon > 0$ there exists a number n_ε such that $|g_{in}(z) - z| < \varepsilon$ for all $z \in D$. If ε is smaller than the difference of the radii of D and D_0, by the argument principle, $D_0 \subset g_{in}(D)$. Therefore, every $z \in D_0$ has an inverse $z_n = f_{in}(z) \in D$. We have

$$|z_n - z| = |f_{in}(z) - z| = |f_{in}(z) - g_{in}(f_{in}(z))| = |z_n - g_{in}(z_n)| < \varepsilon$$

for all $z \in D_0$. We conclude that the inverse mapping f_{in} is defined on D_0 and the sequence (f_{in}) tends to the identity uniformly on D_0. Now

$$z_n = f_{in}(z) = e^{\frac{2\pi i}{a_{in}} \Phi_n(z)}.$$

Therefore $dz_n = z_n \cdot \dfrac{2\pi i}{a_{in}} \Phi'_n(z)\, dz$, and by squaring

$$\varphi_n(z) = -\left(\frac{a_{in}}{2\pi}\right)^2 \frac{1}{z_n^2} \left(\frac{dz_n}{dz}\right)^2.$$

Clearly $\dfrac{dz_n}{dz} \to 1$, and since $\varphi_n(z) \to \varphi(z) \neq 0$ for proper z (φ cannot be the zero differential, because otherwise $a_{in} \to 0$ for all i) we have shown that $a_i = \lim_{n\to\infty} a_{in} \neq 0$ exists for $i = 1, \ldots, q$, and

$$\varphi(z) = -\left(\frac{a_i}{2\pi}\right)^2 \frac{1}{z^2}.$$

This is the representation of the quadratic differential φ in terms of the parameter z. Therefore φ has closed trajectories in R_i, and $\|\varphi\|_{R_i} = a_i^2 M_i$. We finally get

$$\begin{aligned} 1 \geq \|\varphi\| \geq \sum_{i=1}^{q} \|\varphi\|_{R_i} &= \sum_{i=1}^{q} a_i^2 M_i \\ &= \lim_{n\to\infty} \sum_{i=1}^{q} a_{in}^2 \cdot M_{in} = \lim_{n\to\infty} \sum_{i=1}^{p} a_{in}^2 M_{in} = 1. \end{aligned}$$

Therefore φ has closed trajectories on R, by definition.

Let us turn back to the original subsequence (φ_n) which was chosen to converge to φ locally uniformly. From the fact that $\|\varphi\| = 1$ it easily follows that $\|\varphi_n - \varphi\| \to 0$. We have in fact shown that any sequence of quadratic differentials $\varphi_n \in \Gamma_0$ which converges locally uniformly to a quadratic differential φ converges in norm to φ and $\varphi \in \Gamma_0$. Moreover: $M_{in} \to M_i$ and $b_{in} \to b_i$ for all $i = 1, \ldots, p$, and $a_{in} \to a_i$ if $M_i \neq 0$. If this statement were not true for the moduli, we could pass to another subsequence with different limits $\tilde{M}_i$. But the $\tilde{M}_i$ are the moduli of the ring domains of the limit φ, which are well determined. Let now $M_i \neq 0$ for $i = 1, \ldots, q$, $M_i = 0$ for $i = q+1, \ldots, p$. Then, $a_{in} \to a_i$ for $i = 1, \ldots, q$, and hence $b_{in} = a_{in} M_{in} \to b_i = a_i M_i$, whereas $b_{in} \to 0$ for $i = q+1, \ldots, p$, because the sequences (a_{in}) are bounded.

Corollary 21.2. *Let $\varphi_n \in \Gamma$ for $n=1,2,3,\ldots$, and let $\varphi_n \to \varphi$ locally uniformly. Then $\varphi \in \Gamma$ and the sequence (φ_n) tends to φ in norm. If $\varphi = 0$, $a_{in} \to 0$ and $b_{in} \to 0$ for $n \to \infty$. If $\varphi \neq 0$, $M_{in} \to M_i$, $b_{in} \to b_i$ for all i, and $a_{in} \to a_i$ for all i with $M_i > 0$.*

Proof. From the locally uniform convergence $\varphi_n \to \varphi$ it easily follows that

$$\overline{\lim_{n\to\infty}}\, a_{in} \le a_i = \inf_{\gamma \sim \gamma_i} \int_\gamma |\varphi|^{\frac{1}{2}} |dz|.$$

On the other hand, $M_{in} \le \bar{M}_i < \infty$, $i = 1, \ldots, p$. Therefore

$$\overline{\lim_{n\to\infty}}\, \|\varphi_n\| = \overline{\lim_{n\to\infty}} \sum_{i=1}^{n} a_{in}^2 M_{in} \le \sum_{i=1}^{n} a_i^2 \bar{M}_i.$$

If $\varphi = 0$, $a_i = 0$ for all i, and consequently $\|\varphi_n\| \to 0$.

Let $\varphi \neq 0$. We choose a subsequence (φ_{n_i}) such that the sequence $(\|\varphi_{n_i}\|)$ converges. Let $\lambda = \lim_{i\to\infty} \|\varphi_{n_i}\|$. Then, by the previous theorem, the sequence $(\varphi_{n_i}/\|\varphi_{n_i}\|)$ converges in norm to a quadratic differential $\tilde{\varphi} \in \Gamma_0$:

$$\left\| \frac{\varphi_n}{\|\varphi_{n_i}\|} - \tilde{\varphi} \right\| \to 0$$

hence

$$\|\varphi_{n_i} - \lambda \tilde{\varphi}\| \to 0.$$

The sequence (φ_{n_i}) converges, a fortiori, to $\lambda \tilde{\varphi}$ pointwise (locally uniformly), thus $\varphi = \lambda \tilde{\varphi}$, $\|\varphi\| = \lambda \|\tilde{\varphi}\| = \lambda$. Therefore the limit λ is the same for every subsequence, and we have shown that the sequence $(\|\varphi_n\|)$ converges to $\|\varphi\|$, and consequently

$$\|\varphi_n - \varphi\| \to 0.$$

By the previous theorem again, $\varphi = \lambda \tilde{\varphi}$ has closed trajectories, $M_{in} \to M_i$, $b_{in} \to b_i$ for all i, and $a_{in} \to a_i$ for all i with $M_i \neq 0$.

If (φ_n) is merely a bounded sequence of Γ, we can extract a subsequence which converges locally uniformly. The statement of the corollary is then true for this subsequence.

21.3 The homeomorphism $(b_1, \ldots, b_p) \mapsto \varphi$. Let $(\gamma_1, \ldots, \gamma_p)$ be an admissible system of Jordan curves on a Riemann surface R with finite maximal moduli $\bar{M}_i$, $i = 1, \ldots, p$. Then, to every p-tupel of non negative numbers b_i there corresponds a uniquely determined quadratic differential $\varphi \in \Gamma$ with b_i the heights of its cylinders:

$$(b_1, \ldots, b_p) \mapsto \varphi.$$

Of course we assign the zero differential to the zero vector. Conversely, to every $\varphi \in \Gamma$ corresponds the vector of its heights. The mapping is therefore a bijection of the set of non negative p-tupels onto Γ.

Theorem 21.3. *The mapping $(b_1, \ldots, b_p) \mapsto \varphi$ is a homeomorphism (with the metric induced by the norm on Γ).*

Proof. The set of $\varphi \in \Gamma$ with bounded heights is bounded in norm. If not, there would exist a sequence of quadratic differentials $\varphi_n \in \Gamma$ with bounded heights b_{in} and norm $\|\varphi_n\|\to\infty$. Let $\tilde{\varphi}_n=\frac{\varphi_n}{\|\varphi_n\|}$. Then, $\tilde{\varphi}_n\in\Gamma_0$, and there is a subsequence of the sequence $(\tilde{\varphi}_n)$ which converges in norm to an element $\tilde{\varphi}\in\Gamma_0$. For convenience of notation we call the subsequence $(\tilde{\varphi}_n)$ again. Let i be an index with $\tilde{M}_i>0$. Then, $\tilde{b}_{in}=\frac{b_{in}}{\|\varphi_n\|}\to\tilde{b}_i>0$, and therefore $b_{in}\to\infty$ which is a contradiction.

Let now $b_{in}\to b_i$, $i=1,\dots,p$, and let $\varphi_n\in\Gamma$ be the corresponding differentials. Since they are bounded, there is a subsequence (φ_{n_k}) which converges in norm to an element $\varphi\in\Gamma$. The heights b_{in_k} of the cylinders of the φ_{n_k} converge to the heights of the respective cylinders of φ, which are therefore equal to b_i. From the uniqueness of the differential with given heights we conclude that already the original sequence (φ_n) converges to φ locally uniformly and hence, by Corollary 21.2, in norm.

Conversely, let $\varphi_n\to\varphi$ in norm. Then, $\varphi_n\to\varphi$ locally uniformly, and by Corollary 21.2 $b_{in}\to b_i$, $i=1,\dots,p$.

21.4 Exhaustion. The considerations of the preceding sections can be applied to prove the convergence of the quadratic differentials associated with the surfaces of an exhaustion of R.

Theorem 21.4. *Let R be a Riemann surface and let $(\gamma_1,\dots,\gamma_p)$ be an admissible system of Jordan curves on R with maximal moduli $\bar{M}_i<\infty$, $i=1,\dots,p$. For a given, increasing exhaustion $R_1\subset R_2\subset R_3\subset\dots\subset R$ of R and given non negative numbers b_i we denote by φ_n the solution of the heights problem for R_n. Then the differentials φ_n converge locally uniformly to a quadratic differential φ on R. φ has closed trajectories and its cylinders have heights b_i. The convergence is actually in norm: $\|\varphi_n-\varphi\|_{R_n}\to 0$, $M_{in}\to M_i$ and $a_{in}\to a_i$.*

Proof. We may disregard the curves γ_i with associated height $b_i=0$ and thus assume $b_i>0$ for $i=1,\dots,p$. (In particular, if $b_i=0$ for all i, $\varphi_n=0$ for all n, and so is φ. There is nothing to prove.) Without loss of generality we can also assume that the curves γ_i are contained in R_n for all n. Let now φ_n be the differential with cylinders of homotopy type (γ_i) and heights $b_{in}=b_i$ on R_n. Then, by the minimal property and because the system of ring domains $R_{in}\subset R_n$ is contained in R_{n+1},

$$\|\varphi_n\|=\sum_{i=1}^{p}\frac{b_i^2}{M_{in}}\geq\sum_{i=1}^{p}\frac{b_i^2}{M_{i,n+1}}=\|\varphi_{n+1}\|\geq\sum_{i=1}^{p}\frac{b_i^2}{\bar{M}_i}.$$

We conclude that the differentials φ_n have decreasing and thus bounded norm, with a positive lower bound. We can therefore extract a subsequence, which we call φ_n again, which converges locally uniformly to a quadratic differential φ on R, $\|\varphi\|\leq\lim\limits_{n\to\infty}\|\varphi_n\|$. Because of the inequality

$$\frac{b_i^2}{M_{in}}\leq\sum_{i=1}^{p}\frac{b_i^2}{M_{in}}=\|\varphi_n\|\leq\|\varphi_1\|,$$

the moduli M_{in} are bounded away from zero:

$$\frac{b_i^2}{\|\varphi_1\|} \leq M_{in} \leq \bar{M}_i, \qquad i=1, \ldots, p.$$

We can therefore pass to another subsequence (which we denote by (φ_n) again) such that the moduli M_{in} converge,

$$\lim_{n\to\infty} M_{in} = M_i > 0, \qquad i=1, \ldots, p.$$

As in the proof of Theorem 21.2 we introduce the mappings

$$f_{in} = e^{\frac{2\pi i}{a_{in}}\Phi_n}, \qquad \Phi_n = \int \sqrt{\varphi_n}$$

of the ring domains R_{in} onto circular rings R'_{in}: $r_{in} < |z_n| < 1$, $M_{in} = \frac{1}{2\pi} \log \frac{1}{r_{in}}$, and their inverses g_{in}. By passing once more to a subsequence, we can achieve that the sequences g_{in} converge, locally uniformly on R'_i: $r_i < |z| < 1$, $M_i = \frac{1}{2\pi} \log \frac{1}{r_i}$, to conformal mappings g_i of R'_i onto disjoint ring domains R_i on R.

The representation of the differential φ_n in R_i in terms of the parameter z is

$$\varphi_n(z) = -\left(\frac{a_{in}}{2\pi}\right)^2 \frac{1}{z_n^2} \left(\frac{dz_n}{dz}\right)^2, \qquad z_n = f_{in}(z).$$

From the convergence $\varphi_n(z) \to \varphi(z)$ and $f_{in}(z) \to z$ get the representation

$$\varphi(z) = -\left(\frac{a_i}{2\pi}\right)^2 \frac{1}{z^2},$$

with $a_i = \lim_{n\to\infty} = a_{in}$. φ cannot be the zero differential, because otherwise $\overline{\lim_{n\to\infty}}\, a_{in} = 0$ for all i, which contradicts

$$\|\varphi_n\| = \sum_{i=1}^{p} a_{in} b_i \geq \sum \frac{b_i^2}{\bar{M}_i} > 0.$$

Therefore $a_i > 0$ for all i. In R_i, φ has closed trajectories with lengths a_i. The modulus of R_i is $M_i = \frac{1}{2\pi} \log \frac{1}{r_i}$, and its height in the φ-metric is

$$M_i \cdot a_i = \lim_{n\to\infty} M_{in} \cdot a_{in} = \lim_{n\to\infty} b_i = b_i.$$

Moreover,

$$\begin{aligned} \|\varphi\| &\leq \lim_{n\to\infty} \|\varphi_n\| = \lim_{n\to\infty} \sum_{i=1}^{p} \frac{b_i^2}{M_{in}} \\ &= \sum_{i=1}^{p} \frac{b_i^2}{M_i} = \sum_{i=1}^{p} \|\varphi\|_{R_i} \leq \|\varphi\|. \end{aligned}$$

The ring domains R_i of φ cover R up to a set of measure zero, and therefore φ has closed trajectories by definition. The R_i are its characteristic ring domains, in fact φ is the solution of the heights problem for R. From the fact that

$\varphi_n \to \varphi$ locally uniformly and $\|\varphi_n\| \to \|\varphi\|$ we conclude again that $\|\varphi_n - \varphi\| \to 0$. The norm can either be taken over the surfaces R_n or over the surface R; in the latter case we set $\varphi_n = 0$ on $R \setminus R_n$.

The sequence (φ_n) in the last relation is actually a subsequence of the original one. But if the original sequence would not converge to φ locally uniformly there would exist a subsequence which converges, again locally uniformly (because of the local boundedness), to a limit $\tilde{\varphi} \neq \varphi$. The limit $\tilde{\varphi}$ would also solve the heights problem on R, with the same curves γ_i and heights b_i, which contradicts the uniqueness of the solution. Therefore the original sequence tends to φ locally uniformly. Moreover, since the sequence of norms $\|\varphi_n\|$ tends to $\|\varphi\|$, we have in fact $\|\varphi_n - \varphi\| \to 0$ for the original sequence, which finishes the proof of the theorem.

21.5 An existence proof based on the compact case. In the proof of Theorem 21.4 the existence of a solution of the heights problem on the surface R is not assumed; in the contrary, it is proved, if the solutions φ_n on the exhausting surfaces R_n exist. This allows a proof of the existence of a solution on an arbitrary surface if the problem is solved on the compact surfaces. One first solves it on compact bordered surfaces, by completing them with their symmetric image to a compact surface. Then one exhausts R by compact bordered surfaces. The argument reveals that the compact case is the really basic one, for finite curve systems.

Let R be a compact, bordered surface, with boundary curves ∂R_j, $j = 1, 2, \ldots, q$. Let $(\gamma_1, \ldots, \gamma_p)$ denote an admissible system of Jordan curves in the interior of R, with associated heights $b_i > 0$. The mirror image of R is denoted by R^*. $\hat{R} = R \cup R^*$, with corresponding boundary points identified, is the double of R. The symmetric image of γ_i is denoted by γ_i^*. Whenever γ_i is homotopic to some ∂R_j on R, we have $\gamma_i^* \sim \partial R_j^* = \partial R_j$ on R^*, hence $\gamma_i \sim \gamma_i^*$ on $\hat{R}$. In this case we disregard γ_i^* (and may of course replace γ_i by ∂R_j).

We claim that the remaining curve system γ_i, γ_k^* is admissible on $\hat{R}$. We have to show that $\gamma_i \nsim \gamma_k$ for $i \neq k$, and that $\gamma_i \nsim \gamma_k^*$ for all i and k (if there is a γ_k^*).

Let $\gamma_i \sim \gamma_k$ on $\hat{R}$, $i \neq k$. Let D be a ring domain bounded by the two curves. D must have points in common with R^*, because otherwise γ_i and γ_k would be homotopic on R. But as the boundary of D lies in R, D contains all of R^*. R^* must be a subannulus of D, bounded by two curves ∂R_1 and ∂R_2, say. Then, R is an annulus (the symmetric image of R^*), hence $\gamma_i \sim \gamma_k$ on R, a contradiction.

Let now $\gamma_i \sim \gamma_k^*$, and let D be the ring domain bounded by γ_i and γ_k^*. D contains points of R as well as of R^*, hence a boundary curve ∂R_j. This curve cannot be homotopic to zero and must therefore be homotopic to both boundary curves of D. We have $\gamma_i \sim \partial R_j$. But then, γ_i^* was dropped, hence $i \neq k$. As $\gamma_i \sim \gamma_i^*$, $\gamma_i^* \sim \gamma_k^*$, therefore (by reflection) $\gamma_i \sim \gamma_k$ on $\hat{R}$, which was proved impossible.

We now assign heights to the curves of our system: $\gamma_i \to b_i$ as before, unless $\gamma_i \sim \partial R_j$, in which case we set $\gamma_i \to 2b_i$, and $\gamma_k^* \to b_k$. Let $\hat{\varphi}$ be the solution of the heights problem on $\hat{R}$ for the curve system γ_i, γ_k^*, with the assigned heights. Let R_i, R_k^* be the corresponding ring domains. The symmetry T of $\hat{R}$ takes the system of curves (up to homotopy) and heights into itself, hence, by the uniqueness theorem, also the solution $\hat{\varphi}$ and its system of ring domains.

Let R_i be the ring domain associated with γ_i. First, assume $\gamma_i \nsim \partial R_j$ for all j. We want to show that $R_i \subset R$. Let $R_i \cap R^* \neq \emptyset$. If $R_i \subset R^*$, pick a closed trajectory $\alpha_i \subset R_i$. As it does not meet γ_i, it bounds, together with γ_i, a ring domain D, which, in its turn, evidently must contain a boundary curve ∂R_j. Thus $\gamma_i \sim \partial R_j$, in contradiction to the hypothesis. So let R_i contain points of R and of R^*, hence a boundary point P. The symmetry T leaves P invariant and takes R_i into R_i^*. Therefore $R_i \cap R_i^* \neq \emptyset$, which is impossible, unless $R_i = R_i^*$. But then $\gamma_i \sim \gamma_i^*$, a contradiction. We conclude: for every curve γ_i which is not homotopic to a boundary curve ∂R_j the corresponding ring domain is contained in R.

Second, assume $\gamma_i \sim \partial R_j$. Then R_i must be symmetric, i.e. invariant under T. T is therefore an anti-conformal selfmapping of R_i. In a representation of R_i as a circular annulus, the corresponding selfmapping has either a couple of opposite radii or a concentric circle as locus of fixed points. By homotopy reasons the first is impossible. The boundary curve ∂R_j is a trajectory in R_i which bisects R_i in two cylinders of equal height b_i. The restriction of the ring domains to R solves the given heights problem for R.

The boundary curves ∂R_j are either closed trajectories, or else they are composed of critical trajectories which lie on the boundary of characteristic ring domains of φ.

Just as an example we indicate the solution of the heights problem on a compact surface with punctures: We cut little circular holes (in fixed local parameters) around the punctures and exhaust it by letting the radii shrink to zero. The requirement that $\bar{M}_i < \infty$ means that no γ_i bounds a once punctured disk on R.

21.6 The surface of the squares of the heights. Let $(\gamma_1, \ldots, \gamma_p)$ be an admissible system of Jordan curves on an arbitrary Riemann surface R, satisfying the condition $\bar{M}_i < \infty$ for $i = 1, \ldots, p$. The set of normed quadratic differentials φ with closed trajectories and with characteristic ring domains R_i of homotopy type γ_i, $i = 1, \ldots, p$, is denoted by Γ_0. The fact that

$$\|\varphi\| = \sum_{i=1}^{p} \frac{b_i^2}{M_i} = 1$$

is minimized by the ring domains R_i suggests to consider the sum as scalar product of the two vectors $x = (x_1, \ldots, x_p)$, $y = (y_1, \ldots, y_p)$ with $x_i = b_i^2$, $y_i = M_i^{-1}$, $i = 1, \ldots, p$. This gives rise to a concave surface, described by the vector x, with a continuous normal vector y (except for the boundary points with more than one x_i equal to zero).

We first show that in every direction $e = (e_1, \ldots, e_p)$,

$$\|e\| = \left(\sum_{i=1}^{p} e_i^2\right)^{1/2} = 1, \quad e_i \geq 0 \text{ for } i = 1, \ldots, p,$$

there is exactly one vector $x = \lambda \cdot e$, $\lambda > 0$, with $x_i = b_i^2$, b_i the heights of the cylinders of a quadratic differential $\varphi \in \Gamma_0$.

Assume $x=\lambda e$ solves the problem. Then there exists $\varphi\in\Gamma_0$ with heights of its cylinders being $b_i=\sqrt{\lambda}\cdot\sqrt{e_i}$, i.e. multiples of the square roots $\sqrt{e_i}$. Let now $\tilde{\varphi}\in\Gamma$ be the differential with heights $\tilde{b}_i=\sqrt{e_i}$. Then, as

$$1=\|\varphi\|=\sum\frac{b_i^2}{M_i}=\lambda\sum\frac{e_i}{M_i}=\lambda\,\|\tilde{\varphi}\|,$$

we get $\lambda=\|\tilde{\varphi}\|^{-1}$, $\varphi=\lambda\tilde{\varphi}$, where the sum is extended over those indices i, for which $e_i>0$. Thus, φ has the correct heights, and since $\tilde{\varphi}$ is uniquely determined, so is φ and hence x.

Moreover, x depends continuously on e, because $\tilde{\varphi}$ does, and hence λ. It therefore describes a surface $\mathscr{H}$, the surface of the squares of the heights of the cylinders.

We now show that $\mathscr{H}$ is convex. Let x be an interior point of $\mathscr{H}$ (i.e. $x_i>0$ for all i), and let $x'\neq x$. Then

$$(x',y)=\sum_{i=1}^{p}\frac{b_i'^2}{M_i}>\sum_{(b_i'>0)}\frac{b_i'^2}{M_i'}=(x',y')=1, \tag{1}$$

because of the minimum property of the quadratic differential φ' associated with x' (Theorem 20.5). Subtracting $(x,y)=1$ from the left hand side of (1), we get

$$(x'-x,y)>0, \tag{2}$$

which shows, that the surface $\mathscr{H}$ lies to the right of the plane $\mathscr{E}$ through x with normal y, and that it has only x in common with $\mathscr{E}$.

Assume now that x' is also an interior point of $\mathscr{H}$. Then, the vector y' with components $\dfrac{1}{M_i'}$ is finite and we get, analogously to (2), the inequality.

$$(x-x',y')>0. \tag{3}$$

From the continuity of the vector y at interior points of $\mathscr{H}$ (see Section 21.2) we deduce that $y'=y+\varepsilon$, where $\varepsilon=(\varepsilon_1,\dots,\varepsilon_p)$ is an arbitrary small vector, if x' is sufficiently close to x. We therefore get from (2) and (3), dividing by $\|x'-x\|$,

$$0<\left(\frac{x'-x}{\|x'-x\|},y\right)=\left(\frac{x'-x}{\|x'-x\|},y'-\varepsilon\right)<-\left(\frac{x'-x}{\|x'-x\|},\varepsilon\right)\le\|\varepsilon\|.$$

We conclude that the scalar product

$$\left(\frac{x'-x}{\|x'-x\|},y\right)$$

tends to zero with $x'\to x$ for every interior point of $\mathscr{H}$. Therefore the plane $\mathscr{E}$ is in fact the tangent plane of the surface $\mathscr{H}$ at the point x, and $\mathscr{E}$ depends continuously on x in the interior of $\mathscr{H}$.

Let x be a boundary point of $\mathscr{H}$ with only one component $x_i=b_i^2=0$, whereas $x_k=b_k^2>0$ for $k\neq i$. Then $M_i=0$ and therefore the i-th component of y is $M_i^{-1}=\infty$. Let x' be an interior point of $\mathscr{H}$. The unit normal $\dfrac{y'}{\|y'\|}$ has components

$$\frac{y_k'}{\|y'\|} = \frac{1}{M_k' \sqrt{\left(\frac{1}{M_1'}\right)^2 + \dots + \left(\frac{1}{M_k'}\right)^2 + \dots + \left(\frac{1}{M_p'}\right)^2}}.$$

For $x' \to x$ we have $M_i' \to 0$ and $M_k' \to M_k > 0$ for all $k \neq i$. Therefore

$$\frac{y'}{\|y'\|} \to e^{(i)} = (0, \dots, 1, \dots, 0),$$

with $e^{(i)}$ the unit vector parallel to the i-th axis.

We now show that $e^{(i)}$ is the normal to $\mathcal{H}$ at the boundary point x, $x_i = 0$, $x_k > 0$, $k \neq i$. To this end, let x' be an interior point of $\mathcal{H}$ again. We then have

$$\left(\frac{x'-x}{\|x'-x\|}, e^{(i)}\right) = \frac{b_i'^2}{\|x'-x\|} > 0; \tag{4}$$

and, by (3)

$$\left(\frac{x-x'}{\|x-x'\|}, \frac{y'}{\|y'\|}\right) > 0. \tag{5}$$

Now $\frac{y'}{\|y'\|} = e^{(i)} + \varepsilon$, where ε is an arbitrary small vector for $x' \to x$. Combining (4) and (5) gives

$$0 < \left(\frac{x'-x}{\|x'-x\|}, e^{(i)}\right) = \left(\frac{x'-x}{\|x'-x\|}, \frac{y'}{\|y'\|} - \varepsilon\right) < -\left(\frac{x'-x}{\|x'-x\|}, \varepsilon\right) \leq \|\varepsilon\|,$$

and hence

$$\left(\frac{x'-x}{\|x'-x\|}, e^{(i)}\right) \to 0$$

for $x' \to x$. This is clearly also true, if $x' \in \partial\mathcal{H}$.

At the boundary point x of $\mathcal{H}$ with at least two coordinates equal to zero, $x_i = x_k = 0$ say, there does not exist a continuous unit normal. For, in every neighborhood of x there are points x' with $\frac{y'}{\|y'\|} = e^{(j)}$, $j = i$ and $j = k$, respectively, which are orthogonal to each other.

Let us summarize the results in

Theorem 21.5. *Let Γ_0 be the class of normalized quadratic differentials $\varphi(\|\varphi\| = 1)$ with closed trajectories associated with the admissible curve system $(\gamma_1, \dots, \gamma_p)$. Then the vector $x = (b_1^2, \dots, b_p^2)$ of the squares of the heights of the cylinders of φ describes a concave surface $\mathcal{H}$ with a continuous normal vector $y = \left(\frac{1}{M_1}, \dots, \frac{1}{M_p}\right)$ at the interior points of $\mathcal{H}$. The unit normal $\frac{y}{\|y\|}$ of $\mathcal{H}$ is continuous at the interior points of $\mathcal{H}$ and at those boundary points x which have exactly one component $x_i = 0$. It is equal to the unit vector $e^{(i)}$ of the* i-th *axis at these points.*

21.7 Solution of the moduli problem. The surface $\mathcal{H}$ provides an existence proof for a quadratic differential with given ratio of the moduli by elementary geometric means.

Theorem 21.7 (Strebel [4], [6], [11]). *Let $(\gamma_1, \dots, \gamma_p)$ be a finite admissible curve system on an arbitrary surface R, with finite maximal moduli $\bar{M}_i$ for all $i=1, \dots, p$. Then, for any positive unit vector $m=(m_1, \dots, m_p)$, $m_i>0$, $i=1, \dots, p$, there is a quadratic differential φ on R with closed trajectories of homotopy type γ_i and such that the moduli of its ring domains are $M_i=\lambda m_i$, $i=1, \dots, p$. φ is determined up to a positive constant factor; with the normalization $\|\varphi\|=1$ it is uniquely determined.*

Proof. Let $e=(e_1, \dots, e_p)$ be the unit vector with components proportional to the reciprocals $\frac{1}{m_i}$:

$$e_i=\frac{1}{m_i}\frac{1}{\sqrt{\sum\left(\frac{1}{m_k}\right)^2}}.$$

We push the plane orthogonal to the vector e from the left (the origin) to the surface $\mathscr{H}$. Let x be the first point of contact: $x\in\mathscr{H}\cap\mathscr{E}$ and such that there are no points of $\mathscr{H}$ to the left of $\mathscr{E}$. Then, x is necessarily an interior point of $\mathscr{H}$. For, assume that x is a boundary point with exactly one coordinate equal to zero, say $x_i>0$ for $i=1, \dots, p-1$ and $x_p=0$. We choose $x'\in\mathscr{H}$ with coordinates $x'_i=\tau x_i$, $i=1, \dots, p-1$ and $x'_p=t>0$. As the points of the above form with $\tau=1$ lie on the parallel to the p-th axis through x, which is the normal of $\mathscr{H}$ at x, we must have $0<\tau<1$. Moreover, $\tau\to 1$, for $t\to 0$, and also

$$\left(\frac{x'-x}{\|x'-x\|}, e^{(p)}\right)=\frac{t}{\|x'-x\|}\to 0.$$

From $\|x'-x\|=(1-\tau)\sqrt{x_1^2+\dots+x_{p-1}^2+\left(\frac{t}{1-\tau}\right)^2}$ we then conclude that $\frac{t}{1-\tau}\to 0$.

Therefore

$$(x'-x, e)=e_p\cdot t-(1-\tau)\sum_{i=1}^{p-1} x_i\, e_i$$

becomes negative for sufficiently small positive values of t, and the point $x'\in\mathscr{H}$ lies to the left of the plane $\mathscr{E}$ with normal e, contradicting the choice of $\mathscr{E}$.

If x has more than one coordinate equal to zero,

$$x_i>0, \quad i=1, \dots, q, \quad x_{q+1}=\dots=x_p=0,$$

for some $1\le q<p-1$, we can, by a similar procedure, find a point $x'=(\tau x_1, \dots, \tau x_q, t, 0, \dots, 0)\in\mathscr{H}$ which lies to the left of $\mathscr{E}$.

We have thus shown that x is an interior point of $\mathscr{H}$ (it is of course uniquely determined). Let $\varphi\in\Gamma_0$ be the corresponding quadratic differential. The heights of its cylinders are $b_i=\sqrt{x_i}$. The normal to the surface $\mathscr{H}$ at x is the vector $y=\left(\frac{1}{M_1}, \dots, \frac{1}{M_p}\right)$, with M_i the modulus of the cylinder R_i of homotopy type γ_i. But then, $M_i=\lambda m_i$, which proves the theorem.

The uniqueness of φ follows from Theorem 20.6. For, let $\tilde{\varphi}$ with ring domains $\tilde{R}_i$ and moduli $\tilde{M}_i=\tilde{\lambda} m_i$ be another solution. We may assume $\tilde{\lambda}\ge\lambda$, hence $\frac{\tilde{M}_i}{M_i} \geq 1$ for all i. Then, by Theorem 20.6, $\tilde{R}_i = R_i$, $i = 1, \ldots, p$ and $\tilde{\varphi}$ differs from φ by a positive constant factor, which must be equal to one if $\|\tilde{\varphi}\| = 1$.

21.8 Solution of the moduli problem by exhaustion (Strebel [11]). In Sections 21.4 and 21.5 the quadratic differential with given heights of its cylinders was constructed by exhaustion starting with the solution on compact surfaces. A similar procedure is possible for the solution of the moduli problem on an arbitrary Riemann surface R; hence this problem can also be reduced to the case of compact surfaces.

Let γ_i, $i = 1, \ldots, p$ be an admissible curve system on an arbitrary Riemann surface R with maximal moduli $\bar{M}_i < \infty$ for all i, and let $m = (m_1, \ldots, m_p)$ be a unit vector with positive components. Let (R_n) be an increasing exhaustion of R by compact bordered subsurfaces R_n, $R_1 \supset \gamma_i$ for $i = 1, \ldots, p$. Denote by φ_n the solution on $\hat{R}_n = R_n \cup R_n^*$, where R_n^* is the reflection of R_n on its boundary, with m_i replaced by $2m_i$ for every γ_i which is homotopic to a boundary curve of R_n. By the uniqueness of the solution and the symmetry of $\hat{R}_n$ the restriction $\varphi_n | R_n$ is the solution of the moduli problem $M^{(n)} = (M_{1n}, \ldots, M_{pn}) = \lambda_n \cdot m$ on R_n.

On R_{n+1}, because of the extremum property,

$$\frac{M_{in}}{M_{i,n+1}} = \frac{\lambda_n}{\lambda_{n+1}} \leq 1. \tag{1}$$

Moreover, $M_{in} \leq \bar{M}_i$, $i = 1, \ldots, p$, and hence

$$\frac{M_{i1}}{M_{in}} = \frac{\lambda_1}{\lambda_n} \geq \max_{1 \leq i \leq p} \frac{M_{i1}}{\bar{M}_i}.$$

We conclude that the sequence (λ_n) is monotonically increasing and has a finite limit λ. The same is true for the moduli: $\lim_{n \to \infty} M_{in} = \lambda \cdot m_i = M_i$.

We normalize the solution φ_n on R_n to have norm $\|\varphi_n\| = 1$. By passing to a subsequence we can assume locally uniform convergence to a holomorphic quadratic differential on R. As in Section 21.4 we set

$$f_{in} = e^{\frac{2\pi i}{a_{in}} \Phi_n}.$$

Then $g_{in} = f_{in}^{-1}$ is a conformal mapping of the annulus $r_{in} < |z_n| < 1$ onto the ring domain R_{in}. Taking subsequences again, we can assume that the sequences (g_{in}) converge locally uniformly on $r_i < |z| < 1$, $r_i = \lim_{n \to \infty} r_{in}$, to mappings g_i: $r_i < |z| < 1 \to R_i$, with moduli $M_i = \lim_{n \to \infty} M_{in} = \frac{1}{2\pi} \log \frac{1}{r_i}$. Introducing the parameter z in R_i, we find the representation

$$\varphi(z) = -\left(\frac{a_i}{2\pi}\right)^2 \frac{1}{z^2}, \qquad a_i = \lim_{n \to \infty} a_{in}.$$

This representation and the locally uniform convergence lead to the inequalities

$$1 \geq \|\varphi\| \geq \sum_{i=1}^{p} a_i^2 M_i = \lim_{n \to \infty} \sum a_{in}^2 M_{in} = 1$$

from which we conclude that φ has closed trajectories and that the R_i are its characteristic ring domains, with moduli $M_i=\lambda m_i$. Therefore φ solves the problem for R. Since we can start out with an arbitrary subsequence of the original sequence (φ_n), we have actually shown

Theorem 21.8. *Let R be an arbitrary Riemann surface, with a given admissible curve system γ_i, $i=1,\dots,p$, and finite maximal moduli $\bar{M}_i$, $i=1,\dots,p$. Let $m=(m_1,\dots,m_p)$ be a given unit vector with positive components m_i. Then, for an arbitrary exhaustion (R_n) of R, the solutions φ_n ($\|\varphi_n\|=1$) of the moduli problem $M^{(n)}=\lambda_n\cdot m$ on R_n converge in norm to the normalized solution φ of the moduli problem $M=\lambda m$ on R.*

21.9 The surface of moduli (Strebel, [6], [11]). Given the admissible system of curves γ_i, $i=1,\dots,p$ on an arbitrary Riemann surface with $\bar{M}_i<\infty$ for all i, we associate with each non negative unit vector $m=(m_1,\dots,m_p)$, $m_i\geq 0$, $i=1,\dots,p$, the vector of moduli $M=(M_1,\dots,M_p)$, $M=\lambda m$. Since M does not change when we multiply φ by a positive constant, we can normalize the quadratic differential φ: $\|\varphi\|=1$. The set of all normalized differentials is denoted by $\Gamma_0 =\Gamma_0(\gamma_1,\dots,\gamma_p)$. Note that the closed trajectories of φ are of homotopy type γ_i, but it is not excluded that certain ring domains R_i are degenerated (missing), which of course means the same as $M_i=0$.

The uniqueness of M for given m follows from the extremal property (Theorem 20.6). Let, by appropriate renumbering, $m_i>0$, $i=1,\dots,q$, and $m_i=0$, $i =q+1,\dots,p$. And let $M=\lambda m$ and $\tilde{M}=\tilde{\lambda}m$ be two solutions. Then, $M_i=\tilde{M}_i=0$, $i=q+1,\dots,p$. For $1\leq i\leq q$ we have $M_i>0$, $\tilde{M}_i>0$, with quadratic differentials φ and $\tilde{\varphi}$ respectively. By the minimax property

$$\min_{1\leq i\leq q}\left\{\frac{\tilde{M}_i}{M_i}\right\}=\frac{\tilde{\lambda}}{\lambda}\leq 1.$$

Interchanging φ and $\tilde{\varphi}$, the opposite inequality holds too. We therefore have $\lambda =\tilde{\lambda}$, $\tilde{M}_i=M_i$, $i=1,\dots,q$, and thus $M=\tilde{M}$, $\varphi=\tilde{\varphi}$.

The vector of moduli M is a function of the direction m, $M=\lambda m$. We denote the graph of this function by $\mathscr{M}$ and call it the surface of moduli associated with the system of loops γ_i.

The continuity of M follows from Section 21.2. Let

$$m^{(n)}=(m_1^{(n)},\dots,m_p^{(n)}),\ \|m^{(n)}\|=1,\qquad m_i^{(n)}\geq 0,\ i=1,\dots,p,$$

tend to the unit vector m for $n\to\infty$. Let φ_n and φ be the quadratic differentials in Γ_0 with moduli vector $M^{(n)}=\lambda_n\cdot m^{(n)}$, $M=\lambda m$ respectively. By Corollary 21.1 the sequence (φ_n) has a subsequence (φ_{n_k}) which converges in norm to a quadratic differential $\tilde{\varphi}\in\Gamma_0$. Also $(M^{(n_k)})\to\tilde{M}'$ and therefore $\tilde{M}=\tilde{\lambda}\cdot m$. By the uniqueness we conclude $\tilde{M}=M$. Since we can start with an arbitrary subsequence of $(M^{(n)})$, we have shown that actually the original sequence itself converges to M, and $\varphi_n\to\varphi$ in norm.

From the maximum property of the sum $\sum a_i^2 M_i$ follows the convexity of $\mathcal{M}$. To shorten the writing we introduce the notations

$$x=(x_1, \ldots, x_p)=(M_1, \ldots, M_p), \qquad y=(y_1, \ldots, y_p)=(a_1^2, \ldots, a_p^2),$$

with

$$a_i = \inf_{\gamma \sim \gamma_i} \int_\gamma |\varphi(z)|^{\frac{1}{2}} |dz|,$$

the infimum taken over the rectifiable simple loops γ homotopic to γ_i. (If $M_i>0$, a_i is the φ-length of a closed trajectory of φ in R_i.)

Let x be an interior point, $x' \neq x$ an arbitrary point of $\mathcal{M}$. Then, by Theorem 20.4 we have

$$(x', y) = \sum_{i=1}^{p} M_i' a_i^2 < \sum_{i=1}^{p} M_i a_i^2 = (x, y) = 1 \tag{1}$$

Thus

$$(x'-x, y) < 0, \tag{2}$$

which shows that the point x' of $\mathcal{M}$ lies to the left of the plane $\mathscr{E}$ through x and perpendicular to y. Let now $x' \to x$. Then, as $\varphi' \to \varphi$, by Corollary 21.2 again, we have $y' \to y$. Setting $y' = y + \varepsilon$, $e = \dfrac{x'-x}{\|x'-x\|}$, we get from (2)

$$(e, y) < 0 \tag{3}$$

and, interchanging x and x' in (2),

$$(-e, y') = -(e, y+\varepsilon) = -(e, y) - (e, \varepsilon) < 0. \tag{4}$$

From (3) and (4) follows

$$-(e, \varepsilon) < (e, y) < 0, \tag{5}$$

where $|(e, \varepsilon)| \leq \|\varepsilon\| \to 0$ for $x' \to x$. We conclude that

$$\lim_{x' \to x} \left(\frac{x'-x}{\|x'-x\|}, y \right) = 0 \tag{6}$$

for every interior point x of $\mathcal{M}$. Therefore, y is the normal of the surface $\mathcal{M}$, and we have shown that $\mathcal{M}$ has a continuous normal $y=(a_1^2, \ldots, a_p^2)$ at its interior points.

Our next step is to show that y is also the uniquely determined normal at the boundary points of $\mathcal{M}$. Let, for convenience of notation, $x=(x_1, \ldots, x_p)$ be a boundary point, with $x_i > 0$ for $i=1, \ldots, q$, $x_i=0$ for $i=q+1, \ldots, p$, for some q, $1 \leq q \leq p-1$. For $x' \to x$ we have, by Corollary 21.2, $a_i' \to a_i$, $i=1, \ldots, q$. It is conjectured that this is true for all i, as in the case of heights (Section 24.7). But at this point we only need the easy half of it.

Lemma 21.9. *Let γ be a closed curve on an arbitrary Riemann surface R, and let (φ_n) be a sequence of holomorphic quadratic differentials on R tending locally uniformly to φ. Then*

$$\overline{\lim_{n\to\infty}}\, a_n \leq a,$$

with

$$a_n = \inf_{\tilde{\gamma}\sim\gamma} \int_{\tilde{\gamma}} |\varphi_n(z)|^{\frac{1}{2}} |dz|,$$
$$a = \inf_{\tilde{\gamma}\sim\gamma} \int_{\tilde{\gamma}} |\varphi(z)|^{\frac{1}{2}} |dz|,$$

and the infimum taken over all rectifiable loops $\tilde{\gamma}$ in the free homotopy class of γ.

Proof. For given $\varepsilon>0$ choose a piecewise smooth loop $\gamma_1\sim\gamma$ such that

$$\int_{\gamma_1} |\varphi(z)|^{\frac{1}{2}} |dz| < a+\varepsilon.$$

Cover γ_1 by finitely many parameter neighborhoods. Since $\varphi_n(z)\to\varphi(z)$ uniformly on γ_1 in terms of these parameters,

$$\int_{\gamma_1} |\varphi_n(z)|^{\frac{1}{2}} |dz| \to \int_{\gamma_1} |\varphi(z)|^{\frac{1}{2}} |dz|.$$

As

$$a_n \leq \int_{\gamma_1} |\varphi_n(z)|^{\frac{1}{2}} |dz|, \qquad \overline{\lim_{n\to\infty}}\, a_n \leq a+\varepsilon$$

for every $\varepsilon>0$. We thus get the desired result.

Let $x'\in\mathcal{M}$ be an arbitrary point in the neighborhood of x. We again have inequality (1), and hence (2) and (3). Thus

$$\overline{\lim_{x'\to x}}\,(e, y) \leq 0. \tag{7}$$

On the other hand, interchanging x' and x, we get $(-e, y')<0$, and hence

$$\begin{aligned}(e, y) &= (e, y') + (e, y-y') > (e, y-y') \\ &= \sum_{i=1}^{q} e_i(a_i^2 - a_i'^2) + \sum_{i=q+1}^{p} e_i(a_i^2 - a_i'^2).\end{aligned} \tag{8}$$

In the first sum we have $a_i'\to a_i$ and $|e_i|\leq 1$. Therefore, for $x'\to x$, this term goes to zero. In the second term we have $1\geq e_i = \dfrac{x_i'}{\|x'-x\|}\geq 0$. Moreover

$$\underline{\lim_{x'\to x}}\,(a_i^2 - a_i'^2) = a_i^2 - \overline{\lim_{x'\to x}}\, a_i'^2 \geq 0.$$

We conclude that

$$\underline{\lim_{x'\to x}}\, e_i(a_i^2 - a_i'^2) \geq 0, \qquad i=q+1, \ldots, p$$

and hence

$$\underline{\lim_{x'\to x}}\,(e, y) \geq \sum_{i=q+1}^{p} \underline{\lim_{x'\to x}}\, e_i(a_i^2 - a_i'^2) \geq 0. \tag{9}$$

Both inequalities (7) and (9) give the desired result

$$(10) \qquad 0 \le \underline{\lim}_{x' \to x} \left(\frac{x'-x}{\|x'-x\|}, y \right) \le \overline{\lim}_{x' \to x} \left(\frac{x'-x}{\|x'-x\|}, y \right) \le 0,$$

hence

$$\lim_{x' \to x} \left(\frac{x'-x}{\|x'-x\|}, y \right) = 0$$

for all $x \in \mathcal{M}$. The vector y is thus the normal of the surface $\mathcal{M}$ at the point x, and it follows from the freedom of the vector $x' \in \mathcal{M}$ that the normal is uniquely determined. From Theorem 20.4 we conclude that the surface $\mathcal{M}$ has only x in common with the plane $(x'-x, y) = 0$. Alltogether we have proved the following

Theorem 21.9. *Let γ_i, $i=1,\dots,p$, be an admissible system of simple loops on an arbitrary Riemann surface R with finite maximal moduli $\bar{M}_i$. Let $\Gamma_0 = \Gamma_0(\gamma_1,\dots,\gamma_p)$ be the set of normalized holomorphic quadratic differentials φ with closed trajectories of homotopy type γ_i, $i=1,\dots,p$. Then the vector of moduli $M=(M_1,\dots,M_p)$ of the characteristic ring domains of φ describes, as a function $M=\lambda m$ of the direction $m=(m_1,\dots,m_p)$, $m_i \ge 0$, a convex surface $\mathcal{M}$ with normal vector $y=(a_1^2,\dots,a_p^2)$, where a_i is the infimum of the φ-lengths of the loops freely homotopic to γ_i. For those i for which M_i is positive, a_i is the φ-length of the closed trajectories homotopic to γ_i. The normal vector y is continuous at the interior points of $\mathcal{M}$.*

21.10 Maximal weighted sum of moduli (Jenkins [2], Jenkins and Suita [1], Strebel [6], [11]). The maximal property 20.4 of a quadratic differential with closed trajectories calls for the solution of the following extremal problem. To every curve γ_i of an admissible curve system we assign a positive number A_i. Then the sum

$$\sum A_i^2 \tilde{M}_i$$

should be maximized over all systems of disjoint ring domains $(\tilde{R}_i)$ of homotopy type (γ_i), with $\tilde{M}_i$ the modulus of $\tilde{R}_i$. The solution is given by the following

Theorem 21.10. *Let (γ_i), $i=1,\dots,p$ be an admissible curve system on an arbitrary Riemann surface R with finite maximal moduli $\bar{M}_i$. Let A_i be given positive numbers. Then, among all systems of disjoint ring domains $(\tilde{R}_i)$ of homotopy type (γ_i) there is a system (R_i) with maximal sum*

$$\sum A_i^2 M_i, \qquad M_i = \operatorname{mod.}(R_i).$$

The R_i are the characteristic ring domains of a uniquely determined quadratic differential φ with closed trajectories and hence themselves uniquely determined except for the torus.

If $M_i > 0$, A_i is equal to the length a_i of the closed trajectories homotopic to γ_i; if $M_i = 0$, $A_i \le a_i = \inf_{\gamma \sim \gamma_i} \int_\gamma |\varphi(z)|^{\frac{1}{2}} |dz|$.

Proof. We first prove the uniqueness of φ by the last property. Let φ and $\tilde{\varphi}$ be two quadratic differentials, both in $\Gamma(\gamma_1, \ldots, \gamma_p)$, with $a_i = A_i$ if $M_i > 0$, $a_i \geq A_i$ if $M_i = 0$, and analogously, $\tilde{a}_k = A_k$ if $\tilde{M}_k > 0$, $\tilde{a}_k \geq A_k$, if $\tilde{M}_k = 0$. Then, by the maximum property and by the above

$$\|\varphi\| = \sum a_i^2 M_i \geq \sum a_i^2 \tilde{M}_i \geq \sum A_i^2 \tilde{M}_i = \sum \tilde{a}_i^2 \tilde{M}_i = \|\tilde{\varphi}\|.$$

Starting with $\tilde{\varphi}$ instead of φ we get $\|\tilde{\varphi}\| \geq \|\varphi\|$. We therefore must have equality and, consequently,

$$\sum a_i^2 M_i = \sum a_i^2 \tilde{M}_i.$$

By the uniqueness part of Theorem 20.4, $\varphi = \tilde{\varphi}$ and, if R is not the torus, $\tilde{R}_i = R_i$ for all i.

The existence is evident, if $p=1$. For, in this case, all the differentials with closed trajectories homotopic to γ_1 are positive multiples of a single one, φ, say. Then, $\frac{A_1^2}{a_1^2} \cdot \varphi$ has closed trajectories of length A_1, and, by Theorem 20.4, its characteristic ring domain R_1 with modulus M_1, solves the problem.

For $p \geq 2$ we can pick $\varphi \in \Gamma(\gamma_1, \ldots, \gamma_p)$ by means of the surface of moduli.

Let $Y = (Y_1, \ldots, Y_p)$ be the vector with $Y_i = A_i^2$. Let $\mathcal{M}$ be the surface of moduli associated with Γ. We push a plane $\mathcal{E}$ with normal Y from the right (backwards on Y) to the surface $\mathcal{M}$. Let $x = (x_1, \ldots, x_p) = (M_1, \ldots, M_p)$ be a point of contact. Assume that $x_i = M_i > 0$ for $i = 1, \ldots, q$ and $x_i = M_i = 0$ for $i = q+1, \ldots, p$, where q is some fixed number, $1 \leq q \leq p$. The surface $\mathcal{M}$ has a normal $y = (y_1, \ldots, y_p) = (a_1^2, \ldots, a_p^2)$ at x. We then have $(x' - x, Y) \leq 0$ for all $x' \in \mathcal{M}$, and hence also

$$\left(\frac{x' - x}{\|x' - x\|}, Y\right) \leq 0.$$

On the other hand,

$$\left(\frac{x' - x}{\|x' - x\|}, y\right) \to 0$$

for $x' \to x$. As every unit vector $\bar{e} = (\bar{e}_1, \ldots, \bar{e}_p)$ which is orthogonal to y and such that $\bar{e}_i \geq 0$, $i = q+1, \ldots, p$ can be approximated by vectors $\frac{x' - x}{\|x' - x\|}$, $x' \in \mathcal{M}$, we have $(\bar{e}, Y) \leq 0$, $(\bar{e}, y) = 0$ for all such vectors $\bar{e}$. Choosing $\lambda > 0$ such that $Y_1 = \lambda y_1$, we find that

$$\bar{e}_2(Y_2 - \lambda y_2) + \ldots + \bar{e}_q(Y_q - \lambda y_q) + \ldots + \bar{e}_p(Y_p - \lambda y_p) \leq 0.$$

We therefore have

$$\begin{aligned} Y_i - \lambda y_i &= 0, \qquad i = 1, \ldots, p \\ Y_i - \lambda y_i &\leq 0, \qquad i = q+1, \ldots, p. \end{aligned}$$

Let $\varphi \in \Gamma_0$ be the quadratic differential corresponding to x, with lengths a_i. Then $y_i = a_i^2$, while $Y_i = A_i^2$. The quadratic differential $\lambda\varphi$ has lengths $\sqrt{\lambda} \cdot a_i$, which is equal to A_i for $i = 1, \ldots, q$ and $\geq A_i$ for $1 = q+1, \ldots, p$. By the maximum property (Theorem 20.4) its characteristic ring domains R_i are the unique solution of the extremal problem. In particular the point $x \in \mathcal{E} \cap \mathcal{M}$ is uniquely determined.

We have shown here that the solution of the moduli problem involves the solution of the extremal problem for the weighted sum of moduli. The converse is also true, as was shown by Jenkins in [6].

21.11 Direct solution of the problem by variational methods. The theorem proved in the last section consists of two parts. In the first part it says that, given the curves γ_i and the positive numbers A_i, $i=1, \ldots, p$, the ring domains of homotopy type γ_i that maximize the sum

$$\sum_{i=1}^{p} A_i^2 M_i \tag{1}$$

are the characteristic ring domains of a quadratic differential φ with closed trajectories. For the non degenerate ring domains R_i the A_i are the φ-lengths a_i of the corresponding closed trajectories of φ.

The second part is about the degenerate ring domains ($M_i=0$). Then

$$A_i \leq a_i = \inf_{\gamma \sim \gamma_i} \int_{\gamma} |\varphi(z)|^{\frac{1}{2}} |dz|.$$

The first part can be shown directly, by the same variational procedure which was used to show the existence of a φ for given heights of the cylinders (Section 21.1). Let $(R_i)_{i=1,\ldots,p}$ be a system of ring domains of homotopy type (γ_i) which maximizes the sum (1). Because the maximal moduli $\bar{M}_i$ as well as the number p of curves are finite, a normal family argument leads, as before, to such a system (R_i). The only difference is that now, unlike to the former case, some R_i can degenerate. Let, for convenience of notation, $M_i>0, i=1, \ldots, q$, and $M_i=0, i=q+1, \ldots, p$, for some q, $1 \leq q \leq p$. A quasiconformal selfmapping of the surface R takes the non degenerate ring domains R_i into ring domains $\tilde{R}_i$ of the same homotopy type. Let Φ_i be the conformal mapping of the domain R_i, cut radially (with respect to a conformally equivalent circular annulus) onto a horizontal rectangle S_i of the $\zeta=\xi+i\eta$-plane with sides $0<\xi<A_i$, $0<\eta<b_i$ $=A_i \cdot M_i$. As in 21.1, the ring domain $\tilde{R}_i$, cut along the image of the slit of R_i, is mapped conformally onto a horizontal strip $\tilde{S}_i$ of length A_i and height $\tilde{b}_i$ $=A_i \tilde{M}_i$ in the ζ^*-plane (Fig. 63).

The computations are now the same as above, with a_i and $\tilde{a}_i$ replaced by A_i. We get

$$A_i \leq \int_{\alpha} |f_\zeta + f_{\bar{\zeta}}| \, d\xi, \tag{2}$$

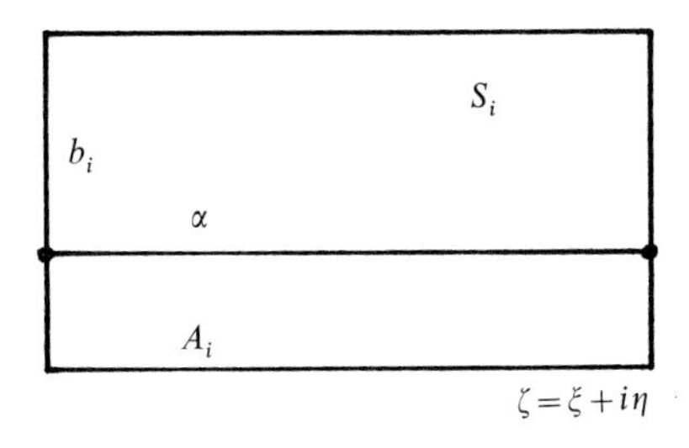

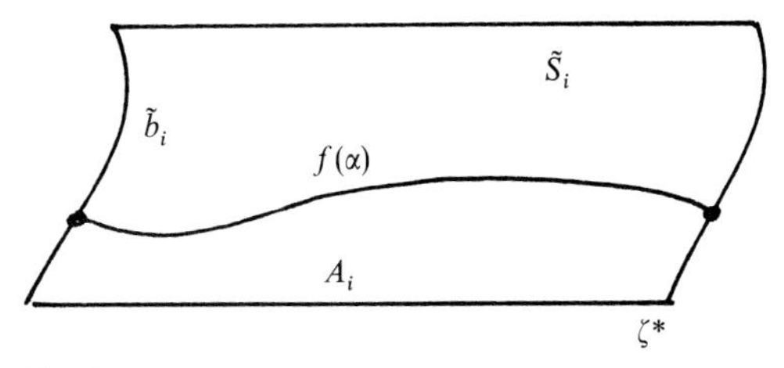

Fig. 63

and after integration over η and summing

(3) $$\sum A_i b_i = \sum A_i^2 M_i \leq \iint_{\cup S_i} |f_\zeta + f_{\bar\zeta}| \, d\xi \, d\eta.$$

An application of Schwarz's inequality leads to

(4) $$\begin{aligned}(\sum A_i^2 M_i)^2 &\leq \iint_{\cup S_i} (|f_\zeta|^2 - |f_{\bar\zeta}|^2) \, d\xi \, d\eta \cdot \iint_{\cup S_i} \frac{|1+\kappa|^2}{1-|\kappa|^2} \, d\xi \, d\eta \\ &= \sum A_i \tilde{b}_i \cdot \iint_{\cup S_i} \frac{|1+\kappa|^2}{1-|\kappa|^2} \, d\xi \, d\eta.\end{aligned}$$

We introduce, as before,

$$\varphi = \begin{cases} \Phi_i'^2 & \text{in } R_i \\ 0 & \text{in } R \setminus \bigcup R_i \end{cases}$$

and write the integral (4) invariantly:

(5) $$(\sum A_i^2 M_i)^2 \leq \sum a_i^2 \tilde{M}_i \cdot \iint_R \frac{|1+\kappa|^2}{1-|\kappa|^2} |\varphi| \, dx \, dy.$$

The quasiconformal selfmapping of R is now specialized, as in Section 21.1, to a shift $z \to w = z + \varepsilon h(z)$ in a neighborhood U and the identity elsewhere. The last integral in (5) then looks like

(6) $$\iint_R \frac{|1+\kappa|^2}{1-|\kappa|^2} |\varphi| \, dx \, dy = \sum A_i^2 M_i + 2 \operatorname{Re} \varepsilon \iint_U h_{\bar z} \varphi \, dx \, dy + O(\varepsilon^2).$$

By construction, the sum $\sum A_i^2 M_i$ is maximal. Therefore $\sum A_i^2 \tilde{M}_i$ can be replaced by $\sum A_i^2 M_i$ in (5), and we get, after dividing (5) by this sum and using (6),

(7) $$0 \leq \operatorname{Re} \varepsilon \iint_U h_{\bar z} \varphi \, dx \, dy + O(\varepsilon^2).$$

This is only possible if $\iint_U h_{\bar z} \varphi \, dx \, dy = 0$. The rest of the proof is the same as in 21.1. From $\Phi_i'^2 = \varphi$ in R_i we conclude that the length of the rectangle S_i is equal to the φ-length of the closed trajectories in R_i: $A_i = a_i$, $i = 1, \ldots, q$. We have thus shown the following

Theorem 21.11. *Let $(\gamma_i)_{i=1,\ldots,p}$ be an admissible curve system, with finite maximal moduli $\bar{M}_i$, on a Riemann surface R, and let A_i be given positive numbers. Let (R_i) be an admissible system of ring domains which maximizes the sum (1). Then the R_i with $M_i > 0$ are the characteristic ring domains of a quadratic differential φ with closed trajectories. Again, for $M_i > 0$, $A_i = a_i$ is the φ-length of the closed trajectories in R_i. Moreover, $b_i = A_i \cdot M_i$, $\|\varphi\| = \sum a_i b_i = \sum A_i^2 M_i$.*

This theorem was shown by Renelt in [1] by a very simple argument, reducing it to Theorem 21.1. Let (R_i), $i = 1, \ldots, q$, be the non degenerate ring domains of a maximal system, and let $(\tilde{R}_i)$ be any system of ring domains of the same homotopy type, with $\tilde{M}_i > 0$ for $i = 1, \ldots, q$. We multiply the elementary inequality

$$\tilde{M}_i + \frac{M_i^2}{\tilde{M}_i} \geq 2 M_i \tag{8}$$

by $\dfrac{b_i^2}{M_i^2}$, with $b_i = A_i M_i$, and sum over i. We get

$$\sum \frac{b_i^2}{M_i^2} \tilde{M}_i + \sum \frac{b_i^2}{\tilde{M}_i} \geq 2 \sum \frac{b_i^2}{M_i}. \tag{9}$$

As $A_i = \dfrac{b_i}{M_i}$, the first sum is

$$\sum A_i^2 \tilde{M}_i \leq \sum A_i^2 M_i = \sum \frac{b_i^2}{M_i}, \tag{10}$$

this leads to

$$\sum \frac{b_i^2}{\tilde{M}_i} \geq \sum \frac{b_i^2}{M_i}. \tag{11}$$

Therefore the system $(R_i)_{i=1,\ldots,q}$ minimizes the sum (11) with $b_i = A_i M_i$. By Theorem 21.1 the R_i are the characteristic ring domains of a quadratic differential φ, with moduli M_i and heights b_i. Therefore $A_i = \dfrac{b_i}{M_i} = a_i$.

It is in the nature of the variational method that no new ring domains appear and no old ones disappear, in other words $\tilde{M}_i > 0$ iff $M_i > 0$. In order to show the second part of Theorem 21.1, namely that $A_i \leq a_i$ if $M_i = 0$, one has to choose a variation which produces a ring domain $\tilde{R}_i$ with $\tilde{M}_i > 0$ where R_i was degenerate. This can be done, using the surface of moduli $\mathscr{M}$ associated with the system (γ_i).

Let $M_i > 0$, $i = 1, \ldots, q < p$, whereas $M_j = 0$ for $q < j \leq p$. Choose an arbitrary index j. As in 21.10 we introduce the vectors

$$x = (x_1, \ldots, x_p) = (M_1, \ldots, M_q, 0 \ldots 0)$$
$$y = (y_1, \ldots, y_p) = (a_1^2, \ldots, a_p^2), \qquad Y = (Y_1, \ldots, Y_p) = (A_1^2, \ldots, A_p^2).$$

The point x lies on the boundary of $\mathscr{M}$. Let $\tilde{x} = (\tilde{M}_1, \ldots, \tilde{M}_q, 0, \ldots, \tilde{M}_j, \ldots, 0)$ be a neighboring point of x on $\mathscr{M}$. Since y is the normal to $\mathscr{M}$ at x, we get, with $e = \dfrac{\tilde{x} - x}{\|\tilde{x} - x\|}$, $(e, y) \to 0$ for $\tilde{x} \to x$. On the other hand, from $(\tilde{x}, Y) \leq (x, Y)$, we conclude $(e, Y) \leq 0$. Therefore

$$\overline{\lim_{\tilde{x} \to x}} ((e, Y) - (e, y)) = \overline{\lim_{\tilde{x} \to x}} e_j (A_j^2 - a_j^2) \leq 0. \tag{12}$$

If we now spezialize $\tilde{x}$ to $\tilde{M}_i = \lambda M_i$, $i = 1, \ldots, q$, clearly e_j, which must be positive, does not go to zero for $\tilde{x} \to x$. It then follows that $A_j \leq a_j$. This finishes the new proof of Theorem 21.10. In particular, the uniqueness of φ and hence of the maximizing system R_i follows.

21.12 Solution by exhaustion. Let γ_i and $A_i > 0$, $i = 1, \ldots, p$ be given on R, $\bar{M}_i < \infty$. And let $(R_n)_{n=1,2,\ldots}$ be an exhaustion of R by subsurfaces, $R_n \subset R_{n+1} \subset R$ for all n, $R = \bigcup R_n$. We may assume that $\gamma_i \subset R_1$ for all i.

Theorem 21.12. *Suppose Theorem* 21.10 *has been shown for subsurfaces* R_n, *with the curves* γ_i *and numbers* A_i. *Then the theorem is also true for the surface* R, *and the solutions* φ_n *on* R_n *tend in norm to the solution* φ *on* R. *Moreover,* $M_{in} \to M_i$, $b_{in} \to b_i$ and, if $M_i > 0$, $a_{in} \to a_i$.

Proof. For $m > n$, the surface R_m contains R_n. Therefore the admissible systems of ring domains on R_n are also admissible on R_m. It follows by considering R_m, that

$$\|\varphi_n\| = \sum_{i=1}^{p} A_i^2 M_{in} \leq \sum_{i=1}^{p} A_i^2 M_{im} = \|\varphi_m\| \tag{1}$$

On the other hand, for all n,

$$\|\varphi_n\| = \sum_{i=1}^{p} A_i^2 M_{in} \leq \sum_{i=1}^{p} A_i^2 \bar{M}_i. \tag{2}$$

The sequence of the norms $\|\varphi_n\|$ therefore has a positive limit

$$\lim_{n\to\infty} \|\varphi_n\| = A < \infty. \tag{3}$$

Since $0 \leq M_{in} \leq \bar{M}_i$, we can choose a subsequence, which we denote by (φ_n) again, which converges locally uniformly to a quadratic differential φ on R, and such that the sequences of moduli converge: $M_{in} \to M_i \geq 0$. From (2) and (3) we conclude that $\|\varphi\| \leq A = \sum_{i=1}^{p} A_i^2 M_i$. Therefore not all M_i can vanish. Let $M_i > 0$ for $i = 1, \ldots, q \leq p$, $M_i = 0$ for all other i. Following the proof of Theorem 21.4 we find, again by passing to subsequences, for $i = 1, \ldots, q$, mappings g_i: $r_i < |z| < 1 \to R_i$, with $M_i = \frac{1}{2\pi} \log \frac{1}{r_i}$. The R_i are disjoint ring domains on R, of homotopy type γ_i, and the representation of φ in terms of z in R_i is

$$\varphi(z) = -\left(\frac{A_i}{2\pi}\right)^2 \frac{1}{z^2}. \tag{4}$$

The equation shows that R_i is swept out by closed trajectories of length A_i of φ, that the height of the cylinder R_i is $b_i = A_i M_i$, and $\|\varphi\|_{R_i} = A_i^2 M_i$. From

$$A \geq \|\varphi\| \geq \sum_{i=1}^{q} \|\varphi\|_{R_i} = \sum A_i^2 M_i = A \tag{5}$$

it follows that $R \setminus \bigcup_{i=1}^{q} R_i$ is a null set. φ therefore has closed trajectories, and from $\varphi_n \to \varphi$ locally uniformly and $\|\varphi_n\| \to \|\varphi\|$ we get $\|\varphi_n - \varphi\| \to 0$. The characteristic ring domains R_{in} tend, for $i = 1, \ldots, q$, to the characteristic ring domains of φ, in the sense that the mappings g_{in}: $r_{in} < |z| < 1 \to R_{in}$ tend locally uniformly to the mappings g_i. For $i = 1, \ldots, q$, $M_{in} > 0$ for all sufficiently large n. Therefore the lengths of the closed trajectories are $a_{in} = A_i = a_i$. From $M_{in} \to M_i$ we conclude that $b_{in} = A_i \cdot M_{in} \to b_i = A_i \cdot M_i$. For $i = q+1, \ldots, q$, $M_{in} \to 0$ by assumption, while $\overline{\lim_{n\to\infty}}\, a_{in} \leq a_i$ by Lemma 21.9. We conclude that $b_{in} \to 0$, and from $a_{in} \geq A_i$ that $a_i \geq A_i$. Therefore φ has the desired properties. It is thus uniquely determined by Theorem 21.10. The uniqueness implies that the original sequence itself converges in norm to φ and has the above properties (with exclusion of the torus, where the ring domain can be shifted around).

The surfaces R_n can be chosen to have finitely many simple analytic boundary curves. The solution of the problem on R_n can then be reduced, by doubling, to the solution on a compact surface. Thus, again, the problem on an arbitrary surface with a finite curve system is reducible to the compact case. (Jenkins [2], where $A_i \leq a_i$ is shown for compact surfaces by Hadamard's variational formula; Jenkins and Suita [1].)

21.13 The generalized extremal length problem (Jenkins [2], [3]). The quadratic differential the characteristic ring domains of which maximize the weighted sum of moduli for given weights is a good candidate to solve a closely related extremal metric problem. For a given finite admissible curve system (γ_i), $i=1, \ldots, p$ on an arbitrary Riemann surface R, with finite maximal moduli $\tilde{M}_i$, and for given positive numbers A_i we call the metric $\rho(z)|dz|$ admissible, if

$$\int_\gamma \rho(z)\,|dz| \geq A_i, \qquad i=1, \ldots, p,$$

for all rectifiable loops γ freely homotopic to γ_i. (The condition can be slightly weakened analogously to the weaker condition in 20.3.) A metric ρ which is square integrable over R and minimizes $\iint_R \rho^2(z)\,dx\,dy$ among all admissible metrics is called extremal; the problem itself is called a generalized extremal length problem, or a generalized extremal metric problem.

It follows from a convexity argument that there can be at most one extremal metric; this is also true for $p=\infty$, and has nothing to do with quadratic differentials. Let ρ_1 and ρ_2 be two extremal metrics,

$$m=\iint_R \rho_1^2\,dx\,dy=\iint_R \rho_2^2\,dx\,dy.$$

Then $\rho=\lambda_1\rho_1+\lambda_2\rho_2$, $\lambda_1, \lambda_2 \geq 0$, $\lambda_1+\lambda_2=1$, is admissible, since for $\gamma \sim \gamma_i$

$$\int_\gamma \rho\,|dz|=\lambda_1\int_\gamma \rho_1\,|dz|+\lambda_2\int_\gamma \rho_2\,|dz| \geq (\lambda_1+\lambda_2)\,a_i=a_i.$$

On the other hand,

$$\begin{aligned} m &\leq \iint_R \rho^2\,dx\,dy=\iint_R (\lambda_1\rho_1+\lambda_2\rho_2)^2\,dx\,dy \\ &=\lambda_1^2\iint_R \rho_1^2\,dx\,dy+2\lambda_1\lambda_2\iint_R \rho_1\rho_2\,dx\,dy+\lambda_2^2\iint_R \rho_2^2\,dx\,dy \\ &\leq (\lambda_1^2+\lambda_2^2)\,m+2\lambda_1\lambda_2\,m=(\lambda_1+\lambda_2)^2\cdot m=m. \end{aligned}$$

Therefore, because of Schwarz's inequality, $\rho_2=c\cdot\rho_1$, $c^2=1$, hence $\rho_2=\rho_1=\rho$ a.e.

It follows from Theorem 20.3 that

$$\sum_{i=1}^{p} A_i^2\,\tilde{M}_i \leq \iint_R \rho^2\,dx\,dy$$

for all admissible metrics ρ and all systems of disjoint ring domains $\tilde{R}_i$ of homotopy type γ_i. Let R_i be the maximizing system. According to Theorem 21.10 it is the system of characteristic ring domains of a uniquely determined quadratic differential φ, with closed trajectories. The lengths of its trajectories are $a_i = A_i$ if $M_i > 0$, whereas $a_i \geq A_i$, with $a_i = \inf\limits_{\gamma \sim \gamma_i} \int\limits_{\gamma} |\varphi(z)|^{\frac{1}{2}} |dz|$, if $M_i = 0$. The metric $|\varphi(z)|^{\frac{1}{2}} |dz|$ is therefore admissible, and by Theorem 20.3 it is the (unique) extremal metric.

For the proof of the extremality (Theorem 20.3) we only need the length inequality for those indices i with $M_i > 0$, i.e. where we actually have ring domains. The proof shows at the same time that there really exists an admissible square integrable metric, namely $\rho = |\varphi|^{\frac{1}{2}}$. The existence of φ depended on the condition that $\bar{M}_i < \infty$ for all i. It follows from Theorem 20.3 that this condition is also necessary for the existence of a square integrable admissible metric ρ, if $A_i > 0$.

21.14 The surface of the reciprocals of the moduli. For a given admissible curve system $(\gamma_1, \ldots, \gamma_p)$ we look at the set of holomorphic quadratic differentials φ with closed trajectories of homotopy type (γ_i), with norm $\|\varphi\| = 1$, and positive moduli for all i (the set Γ_0 without its boundary). We introduce the notations

$$x = (x_1, \ldots, x_p) = \left(\frac{1}{M_1}, \ldots, \frac{1}{M_p}\right),$$

$$y = (y_1, \ldots, y_p) = (b_1^2, \ldots, b_p^2).$$

The vector x describes, as a function of the direction $m = (m_1, \ldots, m_p)$, $\|m\| = 1$, $m_i > 0$ for all i, a surface $\mathscr{M}^{-1}$. Its points are the p-tuples of the reciprocals of the moduli. From the preceding it is clear that:

(1) For every strictly positive direction m we have one and only one vector $x = \lambda m$. To find it we have to solve the moduli problem $\dfrac{1}{M_i} = \lambda m_i$, i.e. $M_i = \dfrac{1}{\lambda} \dfrac{1}{m_i}$ for given positive numbers m_i.

(2) The vector x depends continuously on m and so does φ, and thence the a_i and the b_i. This follows from Theorem 21.9 because the vector $\left(\dfrac{1}{m_i}\right)$ depends continuously on m.

(3) Since, for any two positive vectors x and $\tilde{x} \neq x$ and corresponding vectors y and $\tilde{y}$ we have

$$(\tilde{x}, y) = \sum \frac{b_i^2}{\tilde{M}_i} > \sum \frac{b_i^2}{M_i} = (x, y) = 1,$$

$$(x, \tilde{y}) = \sum \frac{\tilde{b}_i^2}{M_i} > \sum \frac{\tilde{b}_i^2}{\tilde{M}_i} = (\tilde{x}, \tilde{y}) = 1$$

the inequalities

$$(\tilde{x} - x, y) > 0, \quad (\tilde{x} - x, \tilde{y}) < 0$$

hold, and hence, with $e = \dfrac{\tilde{x} - x}{\|\tilde{x} - x\|}$, $(e, y) > 0$ and $(e, \tilde{y}) < 0$. Setting $\tilde{y} = y + \varepsilon$ with $\varepsilon \to 0$ for $\tilde{x} \to x$ we get

$$0<(e, y)=(e, \tilde{y}-\varepsilon)=(e, \tilde{y})-(e, \varepsilon)<-(e, \varepsilon) \leq \|\varepsilon\| .$$

Therefore $y=(b_1^2, \ldots, b_p^2)$ is the normal of the surface $\mathscr{M}^{-1}$ at the point x, and by Corollary 21.2 it depends continuously on x.

The surface $\mathscr{M}^{-1}$ is concave. Its boundary consists of the points with at least one M_i equal to zero, hence $\|x\| \rightarrow \infty$ if x tends to the boundary of $\mathscr{M}^{-1}$.

We can now solve the heights problem: For given $b_i>0$, $i=1, \ldots, p$ we push the plane with normal $y=(y_1, \ldots, y_p)=(b_1^2, \ldots, b_p^2)$ from the left to the surface $\mathscr{M}^{-1}$ (Fig. 64). Its point of contact x is necessarily an interior point of $\mathscr{M}^{-1}$ and it determines the moduli and hence the differential $\varphi \in \Gamma_0$. The squares of its heights are multiples of the given b_i^2, and a multiple of φ by a proper positive constant is the solution of the given heights problem.

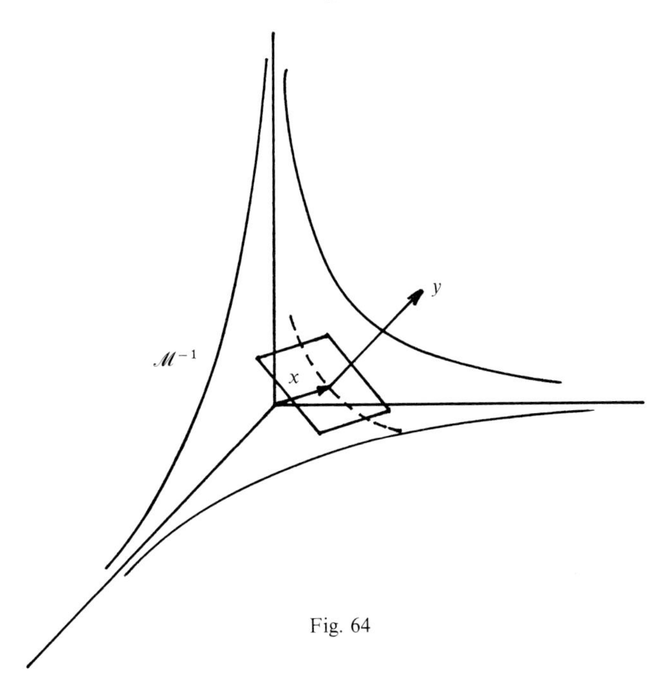

Fig. 64

§ 22. Existence Theorems for Infinite Curve Systems

22.1 Quadratic differentials with given heights of the cylinders. The construction of a quadratic differential with closed trajectories and given heights of its cylinders is, in the case of an infinite admissible system of Jordan curves γ_i, the same as in the case of a finite system. The differential is characterized by its minimal property 20.5: for given positive numbers b_i one has to choose the ring domains R_i of homotopy type γ_i such that

$$\sum \frac{b_i^2}{M_i}=\min .$$

If there is at least one system of ring domains $(\tilde{R}_i)$ of homotopy type (γ_i) such that the corresponding sum

$$\sum \frac{b_i^2}{\tilde{M}_i}, \qquad \tilde{M}_i = \text{mod.}\, \tilde{R}_i$$

is finite, the problem has a unique solution. Conversely, this condition is of course necessary, since for the moduli of the characteristic ring domains R_i of φ the sum $\sum \frac{b_i^2}{M_i}$ is equal to $\|\varphi\|$ which is finite by assumption. Once we have established the existence of a minimizing system of ring domains, the proof goes as before: quasiconformal selfmapping of the surface and application of Weyl's lemma. Since it amounts to an integration over the whole surface, the inconvenience of having infinitely many ring domains is wiped out.

An exhaustion procedure, which consists in adding up more and more curves leads directly to the result. That shows again that the compact case contains the solution of the general problem.

Theorem 22.1 (Renelt [1]). *Let $(\gamma_i)_{i=1,2,3,\ldots}$ be an infinite admissible system of Jordan curves on a Riemann surface R with finite maximal moduli $\bar{M}_i$. Let (b_i) be a sequence of positive numbers which satisfies the following condition: There exists a sequence of disjoint ring domains $\tilde{R}_i$, with moduli $\tilde{M}_i$, $\tilde{R}_i$ of homotopy type γ_i, such that the sum $\sum_i \frac{b_i^2}{\tilde{M}_i}$ is finite.*

Then there exists a uniquely determined quadratic differential φ with closed trajectories, the characteristic ring domains R_i of which are of homotopy type γ_i and have moduli M_i with

$$\sum \frac{b_i^2}{M_i} < \sum \frac{b_i^2}{\tilde{M}_i}$$

for every system $(\tilde{R}_i) \neq (R_i)$. The positive numbers b_i are the heights of the cylinders R_i in the φ-metric, and $\|\varphi\| = \sum \frac{b_i^2}{M_i} < \infty$.

Of course, the ring domains R_i are uniquely determined as the characteristic ring domains of φ, since R is not the torus.

Proof. Let

$$m = \inf_{\{(\tilde{R}_i)\}} \sum_i \frac{b_i^2}{\tilde{M}_i},$$

where the infimum is taken over all systems of disjoint ring domains $(\tilde{R}_i)$ of homotopy type (γ_i). Since there exists, by assumption, a system $(\tilde{R}_i)$ with a finite value of the series $\sum \frac{b_i^2}{\tilde{M}_i}$, m is finite. On the other hand, $\tilde{M}_i \leq \bar{M}_i < \infty$ for all i. Of course, the maximal ring domains with moduli $\bar{M}_i$ are not disjoint, and therefore the sum $\sum_i \frac{b_i^2}{\bar{M}_i}$ is not competing. But from the above we conclude

$$\sum_i \frac{b_i^2}{\bar{M}_i} \leq \sum \frac{b_i^2}{\tilde{M}_i}$$

for all systems (R_i), hence

$$0<\sum_i \frac{b_i^2}{\bar{M}_i} \leq m \leq \sum \frac{b_i^2}{\tilde{M}_i}<\infty .$$

Let $(R_{in})_{i=1,2,\dots}$ for $n=1,2,\dots$ be a minimizing sequence of admissible systems of ring domains. In other words,

$$\lim_{n\to\infty} \sum_i \frac{b_i^2}{M_{in}}=m, \qquad M_{in}=\operatorname{mod.} R_{in}.$$

We denote by g_{in} a conformal mapping of the annulus $r_{in}<|z|<1$, $M_{in} =\frac{1}{2\pi}\log\frac{1}{r_{in}}$, onto R_{in}. Since $\frac{b_i^2}{M_{in}} \leq 2m$ for all sufficiently large n,

$$0<\frac{b_i^2}{2m} \leq M_{in} \leq \bar{M}_i<\infty .$$

Therefore r_{in} is bounded away from zero and one, and the sequence $(g_{in})_{n=1,2,3,\dots}$ is a normal family for every i. We can thus pick a subsequence of the index sequence (n) such that the corresponding subsequence of mappings g_{1n} converges to a mapping g_1 of an annulus $r_1<|z|<1$ onto a ring domain R_1. From this sequence, we pick another subsequence such that the corresponding sequence of mappings g_{2n} converges, a.s.f. Let us denote the diagonal sequence by $(R_{in_\nu})_{\nu=1,2,\dots}$. It is a minimizing sequence and has the property that the sequences $(g_{in_\nu})_{\nu=1,2,\dots}$ of mappings converge, for every i, to mappings g_i of $r_i<|z|<1$ onto ring domains R_i. As $r_{in_\nu}\to r_i$ for $\nu\to\infty$, $M_{in_\nu}\to M_i$. Therefore

$$\sum_{i=1}^{p} \frac{b_i^2}{M_i}=\lim_{\nu\to\infty} \sum_{i=1}^{p} \frac{b_i^2}{M_{in_\nu}} \leq \lim_{\nu\to\infty} \sum \frac{b_i^2}{M_{in_\nu}}=m.$$

We conclude that

$$\sum_{i=1}^{\infty} \frac{b_i^2}{M_i} \leq m. \tag{1}$$

Since the ring domains R_i are disjoint and of homotopy type γ_i, the opposite inequality holds too. We therefore have equality in (1), and the system of ring domains R_i is the desired minimizing system.

We now apply the variational procedure of Section 21.1. The computations for the individual ring domains are the same and the summation in inequality (7) is also valid for infinitely many disjoint domains R_i. From then on, all the integrals are over the surface R or even just over a parameter neighborhood U. The continuation of the proof and the conclusion are the same as before.

It follows from the proof that the numbers b_i are the heights of the cylinders R_i in the φ-metric. Let $\tilde{\varphi}$ be another differential with cylinders $\tilde{R}_i$ of the same homotopy type and with the same heights $\tilde{b}_i=b_i$. Then, by Theorem 20.5, applied to $\tilde{\varphi}$ instead of φ, we have

$$\|\varphi\|=\sum \frac{b_i^2}{M_i} \geq \sum \frac{b_i^2}{\tilde{M}_i}=\|\tilde{\varphi}\|.$$

As $\sum \frac{b_i^2}{M_i}$ is minimal, we must have equality, and therefore $R_i=\tilde{R}_i$ for every i. We conclude that $\tilde{\varphi}$ is a positive multiple of φ, and from $\|\tilde{\varphi}\|=\|\varphi\|$ that $\tilde{\varphi}=\varphi$.

22.2 Existence proof by adding up the curves. It is possible to prove the existence of a quadratic differential for infinite curve systems by adding up more and more curves and making use of the solution for finite systems. Given the curve system (γ_i) and the positive numbers b_i, satisfying the conditions of Theorem 22.1, let φ_p be the solution for the finite curve system $(\gamma_i)_{i=1,\ldots,p}$ with heights b_i. Every φ_p is defined on the whole surface R. We denote its characteristic ring domains by R_{ip}, with moduli M_{ip}, $i=1,\ldots,p$. By the minimal property 20.5 the following inequalities hold for $q>p$

$$\|\varphi_p\| = \sum_{i=1}^{p} \frac{b_i^2}{M_{ip}} < \sum_{i=1}^{p} \frac{b_i^2}{M_{iq}} < \sum_{i=1}^{q} \frac{b_i^2}{M_{iq}} = \|\varphi_q\|. \tag{1}$$

Therefore the norms of the φ_p are monotonically increasing. On the other hand, there exists, by assumption, a system of ring domains $\tilde{R}_i$ with moduli $\tilde{M}_i$ such that $\sum \frac{b_i^2}{\tilde{M}_i} < \infty$. Again by 20.5,

$$\|\varphi_p\| = \sum_{i=1}^{p} \frac{b_i^2}{M_{ip}} < \sum_{i=1}^{p} \frac{b_i^2}{\tilde{M}_i} < \sum_{1}^{\infty} \frac{b_i^2}{\tilde{M}_i}. \tag{2}$$

The sequence $(\|\varphi_p\|)$ is therefore bounded and has a finite limit $A = \lim_{p\to\infty} \|\varphi_p\|$.

We now apply the procedure of Section 21.1. By passing to a subsequence, which we call (φ_p) again, we can assume that (φ_p) converges locally uniformly to a differential φ, $\|\varphi\| \le A$. The mapping $f_{ip} = \exp\left(\frac{2\pi i}{\alpha_{ip}} \Phi_p\right)$ maps the ring domain R_{ip} onto an annulus $r_{ip} < |z_p| < 1$, $1 \le i \le p$, $p = 1, 2, 3, \ldots$. The radii r_{ip} are, for each i, bounded away from zero and one. The inverse mappings g_{ip} therefore form a normal family. We can pass to a subsequence which converges locally uniformly to a conformal mapping of a circular annulus $r_i < |z| < 1$ onto a ring domain R_i of homotopy type γ_i. Starting with $i=1$ and, for every next i, taking a subsequence of the former one, we finally pass to the diagonal sequence. We call it (φ_p) again. It has the property that the mappings g_{ip} converge, locally uniformly, to mappings g_i for every i. By means of g_i we can introduce the parameter z in R_i and, locally, in R_{ip} for all sufficiently large p. Therefore

$$z_p = f_{ip}(z) = \exp\left(\frac{2\pi i}{a_{ip}} \Phi_p(z)\right) \tag{3}$$

makes sense. We have

$$\varphi_p(z) = -\left(\frac{a_{ip}}{2\pi}\right)^2 \frac{1}{z_p^2} \left(\frac{dz_p}{dz}\right)^2, \tag{4}$$

$M_{ip} = \frac{1}{2\pi} \log \frac{1}{r_{ip}} = \frac{b_i}{a_{ip}}$. From $z_p \to z$, locally uniformly, and $\varphi_p(z) \to \varphi(z)$, we conclude that φ has the representation

$$\varphi(z) = -\left(\frac{a_i}{2\pi}\right)^2 \frac{1}{z^2}, \tag{5}$$

$a_i = \lim_{p\to\infty} a_{ip}$, in terms of the parameter z in R_i. This shows that R_i is swept out by closed trajectories of φ, and of course the R_i are disjoint and of homotopy type γ_i.

For given $\varepsilon>0$ we now choose p such that $\|\varphi_p\|>A-\varepsilon$. Letting q tend to infinity in (1), for fixed p, we get

$$A-\varepsilon<\sum_{i=1}^{p}\frac{b_i^2}{M_i}<\sum_{i=1}^{\infty}\frac{b_i^2}{M_i}\leq\|\varphi\|\leq A. \tag{6}$$

We conclude that

$$\|\varphi\|=\sum_{i=1}^{\infty}\frac{b_i^2}{M_i}=A, \tag{7}$$

with M_i the modulus of R_i. This shows that φ has closed trajectories and that the R_i are its characteristic ring domains. Moreover, $\|\varphi_p-\varphi\|\to 0$ for $p\to\infty$, since $\varphi_p\to\varphi$ locally uniformly and $\|\varphi_p\|\to\|\varphi\|$.

By the uniqueness of φ it follows, as usual, that the original sequence itself converges in norm to φ. We have shown

Theorem 22.2. *Let $(\gamma_i)_{i=1,2,\ldots}$ be an infinite admissible system of Jordan curves on an arbitrary Riemann surface R, with finite maximal moduli $\bar{M}_i$ and such that there exists a system of disjoint ring domains $\tilde{R}_i$ of homotopy type γ_i with $\sum\frac{b_i^2}{\tilde{M}_i}<\infty$. Then, the quadratic differentials φ_p, associated with the curve systems $(\gamma_i)_{i=1,\ldots,p}$ and heights b_i, converge in norm to the differential φ, associated with the system $(\gamma_i)_{i=1,2,\ldots}$ and with heights b_i, $i=1,2,\ldots$.*

As a corollary we can note that every differential $\varphi\in\Gamma(\gamma_1,\gamma_2,\ldots)$ is the limit, in norm, of differentials $\varphi_p\in\Gamma(\gamma_1,\gamma_2,\ldots)$ with only a finite number of characteristic ring domains.

Another consequence is the following: In Section 21.5 we showed that, for finite curve systems, the existence proof can be given by means of the solutions on compact surfaces, using reflection and relatively simple normal family arguments. By Theorem 22.2 this is still true for infinite curve systems. The differential φ with infinitely many characteristic ring domains is approximated in norm by a differential φ_p and this, in term, is the limit in norm of the differentials associated with an exhaustion (R_n) of R.

22.3 Weighted sum of moduli: Maximal system of ring domains. In the case of an infinite curve system (γ_i) with assigned positive numbers A_i the finiteness of the maximal moduli $\bar{M}_i$ no longer guarantees the boundedness of the sums $\sum A_i^2\tilde{M}_i$ for all admissible systems of ring domains $\tilde{R}_i$. And even if they are bounded it is not sure that there exists a system (R_i) with maximal sum $\sum A_i^2 M_i$. However, if there is such a system (R_i), Theorem 21.11 holds.

Theorem 22.3. *Let (γ_i) be an infinite admissible system of Jordan curves on an arbitrary Riemann surface R. Let, for given positive numbers A_i, the admissible system of ring domains R_i be maximal in the following sense: For every admissible system $(\tilde{R}_i)$ with $\tilde{M}_i>0$ if and only if $M_i>0$ we have*

$$\sum A_i^2\tilde{M}_i\leq\sum A_i^2 M_i<\infty.$$

Then the non degenerate R_i are the characteristic ring domains of a quadratic differential φ with closed trajectories. The length of the trajectories in R_i is A_i, and $\|\varphi\| = \sum A_i^2 M_i$.

The proof of the theorem is the same as the proof of Theorem 21.11, either by quasiconformal variation of the system (R_i) or by Renelt's remark, using the variational proof of Theorem 22.1.

A "local" uniqueness theorem can also be shown: Let $(\tilde{R}_i)$ be an admissible system of ring domains with $\tilde{M}_i > 0 \Rightarrow M_i > 0$ (i.e. we allow degenerate $\tilde{R}_i$, but if $\tilde{R}_i$ is not degenerate, then φ has a characteristic ring domain R_i). Then

$$\sum A_i^2 \tilde{M}_i < \sum A_i^2 M_i$$

unless $(\tilde{R}_i) = (R_i)$.

The proof is the same as that of Theorem 20.4, using the φ metric. We know that every closed curve in $\tilde{R}_i$ separating its two boundary components has φ-length $\geq A_i$, because the closed trajectories in R_i have φ-length A_i.

We do not know, if $A_i \leq a_i$ for those indices i with $M_i = 0$. However, this follows from another approach of the problem (see Section 22.5). It is then a consequence of Theorem 20.4 that the maximizing system of ring domains is uniquely determined.

In the following section, a sufficient condition is given which guarantees both, the existence and uniqueness of a maximizing system (R_i).

22.4 Exhaustion of the curve system

Theorem 22.4. *Let (γ_i) be an infinite admissible system of Jordan curves on a Riemann surface R, with finite maximal moduli $\bar{M}_i$. Let (A_i) be a sequence of positive numbers such that $\sum A_i^2 \bar{M}_i < \infty$. Then, an admissible system of ring domains R_i with maximal sum*

$$\sum A_i^2 M_i$$

exists and is uniquely determined. It is the system of characteristic ring domains of a uniquely determined quadratic differential φ with closed trajectories, which satisfies $A_i \leq a_i$ for all i, and $A_i = a_i$ if $M_i > 0$.

The solutions φ_p of the corresponding problem for finite curve systems $(\gamma_1, \ldots, \gamma_p)$ converge in norm to φ, and $M_{ip} \to M_i$, $b_{ip} \to b_i$, $a_{ip} \to a_i$ for $M_i > 0$.

Proof. Let φ_p be the solution, on R, for the finite curve system $(\gamma_1, \ldots, \gamma_p)$. Let R_{ip}, $i = 1, \ldots, p$, be its ring domains with moduli $M_{ip} \geq 0$. The sum

$$\sum_{i=1}^{p} A_i^2 M_{ip} \tag{1}$$

is then maximal in the class $\Gamma(\gamma_1, \ldots, \gamma_p)$. For $q > p$ we have

$$\begin{aligned} \|\varphi_p\| = \sum_{i=1}^{p} A_i^2 M_{ip} \leq \sum_{i=1}^{q} A_i^2 M_{iq} = \|\varphi_q\| \leq \sum_{i=1}^{q} A_i^2 \bar{M}_i \\ \leq \sum_{i=1}^{\infty} A_i^2 \bar{M}_i. \end{aligned} \tag{2}$$

Therefore the limit

$$A=\lim_{p\to\infty}\|\varphi_p\|$$

exists. For every i and $p\geq i$ the inequalities $0\leq M_{ip}\leq \bar{M}_i$ hold. By a proper choice of nested subsequences we end up with a diagonal sequence which, for notational convenience, we call (φ_p) again, with the following properties:

(3) $$\varphi_p\to\varphi \qquad \text{locally uniformly}$$

(4) $$M_{ip}\to M_i, \qquad 0\leq M_i\leq \bar{M}_i \text{ for } p\to\infty$$

(5) for every i with $M_i>0$ the mappings $g_{ip}\colon r_{ip}<|z|<1\to R_{ip}$ converge, locally uniformly, to mappings $g_i\colon r_i<|z|<1\to R_i$, where

$$M_{ip}=\frac{1}{2\pi}\log\frac{1}{r_{ip}}, \qquad M_i=\frac{1}{2\pi}\log\frac{1}{r_i}.$$

The representation of φ_p in terms of the parameter z in R_i is, for sufficiently large p (see Section 21.2)

(6) $$\varphi_p(z)=-\left(\frac{A_i}{2\pi}\right)^2\frac{1}{z_p^2}\left(\frac{dz_p}{dz}\right)^2$$

which tends, locally uniformly on both sides, to

(7) $$\varphi(z)=-\left(\frac{A_i}{2\pi}\right)\frac{1}{z^2}, \qquad r_i<|z|<1.$$

The R_i are therefore swept out by closed trajectories of φ of length A_i. The φ-height of the cylinder R_i is $b_i=A_iM_i=\lim_{p\to\infty}A_iM_{ip}=\lim_{p\to\infty}b_{ip}$. For $M_i=0$, $M_{ip}\to 0$ and hence $b_{ip}=A_iM_{ip}\to 0$.

For any $\varepsilon>0$ we choose p such that

(8) $$A-\varepsilon<\|\varphi_p\|$$

and

(9) $$\sum_{i=p+1}^{\infty}A_i^2\bar{M}_i<\varepsilon.$$

Then, for $q>p$, we have

(10) $$\begin{aligned} A-\varepsilon<\|\varphi_q\| &= \sum_{i=1}^{q}A_i^2M_{iq} \\ &= \sum_{i=1}^{p}A_i^2M_{iq}+\sum_{i=p+1}^{q}A_i^2M_{iq}<\sum_{i=1}^{p}A_i^2M_{iq}+\varepsilon. \end{aligned}$$

Let now $q\to\infty$. As $M_{iq}\to M_i$ for $1\leq i\leq p$ we get

(11) $$A-\varepsilon<\sum_{i=1}^{p}A_i^2M_i+\varepsilon.$$

From the locally uniform convergence of the sequence (φ_p) to φ and $\|\varphi\|_{R_i}=A_i^2M_i$ we get, with the help of (11) for $\varepsilon\to 0$,

(12) $$A\geq\|\varphi\|\geq\sum_{i=1}^{\infty}A_i^2M_i\geq A.$$

Therefore $R \setminus \bigcup R_i$ is a null set. The quadratic differential φ has closed trajectories, (R_i) being the set of its characteristic ring domains, and as $A = \|\varphi\|$, $\lim_{p\to\infty} \|\varphi_p - \varphi\| = 0$.

We still have to show that $A_i \leq a_i = \inf_{\gamma \sim \gamma_i} \int_\gamma |\varphi(z)|^{\frac{1}{2}} |dz|$ for those indices i with $M_i = 0$. But this follows, as at the end of Section 21.12, from $\overline{\lim}_{p\to\infty} a_{ip} \leq a_i$ and $A_i \leq a_{ip}$. The first inequality is the upper semicontinuity of the lengths a_{ip} (Lemma 21.9) and the second the assumption about φ_p. We now get the uniqueness of the system (R_i) and thus of φ from Theorem 20.4. With this, we again conclude that already the original sequence (φ_p) converges in norm to φ.

Theorem 22.4 allows the following complement on extremal metrics.

Corollary 22.4. *For a given (infinite) admissible curve system (γ_i) on an arbitrary Riemann surface R let $A_i > 0$ be such that $\sum A_i^2 \bar{M}_i < \infty$. Then, in the set of admissible metrics $\rho(z)|dz|$, $\int_\gamma \rho(z)|dz| \geq A_i$ for $\gamma \sim \gamma_i$, $i = 1, 2, \ldots$, there exists a unique extremal metric. It is the metric $\rho(z)|dz| = |\varphi(z)|^{\frac{1}{2}}|dz|$, where φ is the quadratic differential with closed trajectories the characteristic ring domains of which maximize the sum $\sum A_i^2 M_i$.*

22.5 The extremal metric in the general case. For infinite curve systems (γ_i) the extremal metric problem need not have a solution by a quadratic differential with closed trajectories. However, provided there is an admissible metric with finite square norm, there also exists a unique extremal metric given by a quadratic differential of finite norm.

Theorem 22.5. *Let (γ_i) be an infinite admissible system of Jordan curves on an arbitrary Riemann surface R. Assume that, for given positive numbers A_i assigned to the curves γ_i, there is an admissible metric $\rho|dz|$ of finite square norm (see Section 21.13). Then, there is a unique extremal metric. It is of the form $\rho|dz| = |\varphi(z)|^{\frac{1}{2}}|dz|$, where φ is a holomorphic quadratic differential of finite norm. If φ_n denotes the solution of the extremal length problem for the finitely many curves γ_i, with lengths A_i, $i = 1, \ldots, n$, the sequence φ_n tends in norm to φ: $\|\varphi_n - \varphi\| \to 0$.*

Proof. Let φ_n be the quadratic differential with closed trajectories on R, of type $(\gamma_1, \ldots, \gamma_n)$, which is the solution of the extremal length problem with lengths A_i, $i = 1, \ldots, n$. If we set, as usual, $a_{in} = \inf_{\gamma \sim \gamma_i} \int_\gamma |\varphi_n(z)|^{\frac{1}{2}}|dz|$ we have $a_{in} \geq A_i$ for $i = 1, \ldots, n$, with equality for every i for which φ_n has a non degenerate characteristic ring domain R_{in} (see Section 21.10). Since the metric $\rho|dz|$ is admissible for the finite problem, we have $\|\varphi_n\| \leq \iint_R \rho^2 dx\,dy$. Therefore, the φ_n are locally bounded, and there exists a subsequence (φ_{n_k}) which converges, locally uniformly, to a holomorphic quadratic differential φ, with norm

$$\|\varphi\| \leq \underline{\lim}_{k\to\infty} \|\varphi_{n_k}\|. \tag{1}$$

Suppose γ is a rectifiable closed loop freely homotopic to γ_i. Then, $\varphi_{n_k} \to \varphi$ uniformly on γ. Since, for $n \geq i$ $A_i \leq \int_\gamma |\varphi_n|^{\frac{1}{2}} |dz|$, we conclude that $A_i \leq \int_\gamma |\varphi|^{\frac{1}{2}} |dz|$, and hence

$$A_i \leq a_i = \inf_{\gamma \sim \gamma_i} \int_\gamma |\varphi|^{\frac{1}{2}} |dz|. \tag{2}$$

Therefore the metric $|\varphi|^{\frac{1}{2}} |dz|$ is admissible for the given extremal length problem, in particular for the finite problem for every n. We have

$$\|\varphi\| \leq \underline{\lim}_{k \to \infty} \|\varphi_{n_k}\| \leq \overline{\lim}_{k \to \infty} \|\varphi_{n_k}\| \leq \|\varphi\|. \tag{3}$$

We thus find $\|\varphi_{n_k}\| \to \|\varphi\|$. This, together with the locally uniform convergence $\varphi_{n_k} \to \varphi$ gives $\|\varphi_{n_k} - \varphi\| \to 0$. In particular, φ is an extremal metric for the infinite problem.

Corollary 22.5. *We have equality in relation* (6′) *of Theorem* 20.3. *A maximizing system of ring domains is uniquely determined (except for the torus).*

The proof is an application of Theorem 20.4 to the above differential φ.

§23. Quadratic Differentials with Second Order Poles

23.1 Extremal properties. The inequalities of Sections 20.4 (weighted sum of moduli) and 20.6 (minimax property of moduli) carry over to quadratic differentials which determine punctured disks, if the moduli are replaced by reduced moduli (Teichmüller [1]). Recall that the reduced modulus of a punctured disk with respect to a given local parameter near the puncture turns out to be equal to $\frac{1}{2\pi} \log r$, where r is the mapping radius of the disk (see Section 3.2).

Let φ be a quadratic differential with closed trajectories, the characteristic ring domains of which consist of finitely many punctured disks R_j and possibly infinitely many cylinders R_k with finite moduli M_k. We exclude the doubly punctured sphere and assume that the reduced norm

$$\sum a_j^2 M_j + \sum a_k^2 M_k$$

of φ is finite. Here, M_j denotes the reduced modulus of R_j with respect to some fixed parameter z_j near the pointlike boundary component P_j, $z_j = 0 \leftrightarrow P_j$.

Let now $(\tilde{R}_i)$ be a system of non overlapping ring domains on R of homotopy type (α_i), where (α_i) is a complete sample set of closed trajectories of φ, one out of every characteristic ring domain. We use the index i for both kinds of ring domains without distinction, but j for punctured disks and k for ring domains with finite moduli. We allow $\tilde{R}_k$ to be degenerate (missing). But every $\tilde{R}_j$ is a punctured disk with puncture P_j and reduced modulus $\tilde{M}_j$ with respect to the same fixed parameter z_j. Then

$$\sum a_i^2 \tilde{M}_i \leq \sum a_i^2 M_i \tag{1}$$

with $a_i = \int_{\alpha_i} |\varphi(z)|^{\frac{1}{2}} |dz|$ the φ-length of the trajectories. Equality holds only if $\tilde{R}_i = R_i$ for all i.

Proof. To derive the inequality we introduce a natural parameter ζ_j near the second order poles P_j (see Section 6.3). The representation of φ in terms of such a parameter is

$$\varphi(\zeta_j)\, d\zeta_j^2 = -\left(\frac{a_j}{2\pi}\right)^2 \frac{d\zeta_j^2}{\zeta_j^2}. \tag{2}$$

The inequality (1) itself can be shown by cutting out circular holes $|\zeta_j| \leq \rho_j$, $\rho_j > 0$, around the points P_j. Since the new boundary components of the truncated surface are closed trajectories of φ, we can apply Theorem 20.4 for quadratic differentials of finite norm. Adding $\sum_j a_j^2 \frac{1}{2\pi} \log \rho_j$ on both sides and letting ρ_j tend to zero we get the result.

However, in order to discuss the equality sign, we have to be more accurate. So let the domains $\tilde{R}_j$ be mapped onto punctured disks $0 < |z_j| < \tilde{r}_j$ by schlicht functions with derivatives

$$\frac{dz_j}{d\zeta_j}(P_j) = 1. \tag{3}$$

We choose $\rho_j = \rho$ for all j. If ρ is sufficiently small, the disk $|z_j| \leq \rho$ is contained in $\tilde{R}_j \cap R_j$, for all j. The moduli of the ring domains $\tilde{R}_j(\rho) = \tilde{R}_j \setminus \{|z_j| \leq \rho\}$ and $R_j(\rho) = R_j \setminus \{|z_j| \leq \rho\}$ are denoted by $\tilde{M}_j(\rho)$ and $M_j(\rho)$ respectively. Evidently, $\tilde{M}_j(\rho) = \frac{1}{2\pi} \log \frac{\tilde{r}_j}{\rho}$, and $\tilde{M}_j(\rho) + \frac{1}{2\pi} \log \rho = \frac{1}{2\pi} \log \tilde{r}_j = \tilde{M}_j$, whereas $M_j(\rho) + \frac{1}{2\pi} \log \rho \to M_j$ for $\rho \to 0$. We cut the annulus $\rho < |z_j| < \tilde{r}_j$ along a radius and map it onto a horizontal rectangle with sides a_j, $\tilde{b}_j(\rho) = a_j \cdot \tilde{M}_j(\rho)$ in the $z = x + iy$-plane. Since a horizontal $y = \text{const.}$ corresponds to a closed curve which is freely homotopic to α_j, we have

$$a_j \leq \int |\varphi(x+iy)|^{\frac{1}{2}} dx, \tag{4}$$

and by the Schwarz inequality

$$a_j \leq \int |\varphi(x+iy)|\, dx. \tag{5}$$

Assume that for some j one of the circles $|z_j| = \text{const.}$ is not a trajectory of φ. Then, there are positive numbers ε and δ such that

$$a_j + \varepsilon \leq \int |\varphi(x+iy)|\, dx \tag{6}$$

for all y in the δ-neighborhood of some y_0. We get by integrating over y

$$a_j \cdot \tilde{b}_j(\rho) + \varepsilon \cdot \delta \leq \iint_{\tilde{R}_j(\rho)} |\varphi(z)|\, dx\, dy. \tag{7}$$

Doing the same thing with the ring domains $\tilde{R}_k$ and then summing over the j and the k we get, on the left hand side,

$$\sum_j a_j^2 \tilde{M}_j(\rho) + \sum_k a_k^2 \tilde{M}_k + \varepsilon \cdot \delta \tag{8}$$

whereas the right hand side is majorized by the norm of φ taken over the truncated surface $R \setminus \bigcup_j \{|z_j| \leq \rho\}$. For any R_k we have $\|\varphi\|_{R_k} = a_k^2 M_k$, whereas for $R_j(\rho)$ the norm $\|\varphi\|_{R_j(\rho)}$ is equal to $a_j^2 M_j(\rho)$, up to an additive term which tends to zero with $\rho \to 0$, because of (3). Adding $\sum_j a_j^2 \frac{1}{2\pi} \log \rho$ on both sides and letting $\rho \to 0$ we get

$$\sum_j a_j^2 \tilde{M}_j + \sum_k a_k^2 \tilde{M}_k + \varepsilon\delta \le \sum_j a_j^2 M_j + \sum_k a_k^2 M_k. \tag{9}$$

Without the assumption about the curves $|z_j| = \text{const.}$ after inequality (5) above we do not have the $\varepsilon \cdot \delta$-term: This proves the inequality (1). Assume now that equality holds. Then the circles $|z_j| = \rho$ are actually trajectories of φ. From the equality

$$\sum_j a_j^2 \tilde{M}_j + \sum_k a_k^2 \tilde{M}_k = \sum_j a_j^2 M_j + \sum_k a_k^2 M_k \tag{10}$$

we conclude, subtracting $a_j^2 \frac{1}{2\pi} \log \rho$ on both sides, that

$$\sum_j a_j^2 \tilde{M}_j(\rho) + \sum_k a_k^2 \tilde{M}_k = \sum_j a_j^2 M_j(\rho) + \sum_k a_k^2 M_k. \tag{11}$$

The right hand side is equal to the norm of φ over the truncated surface $R(\rho) = R \setminus \bigcup_j \{|z_j| \le \rho\}$, because φ is a quadratic differential with closed trajectories on $R(\rho)$, and the $R_j(\rho)$ and R_k are its characteristic ring domains. Theorem 20.4 yields $\tilde{R}_j(\rho) = R_j(\rho)$, $\tilde{R}_k = R_k$ for all j and k. Hence also $\tilde{R}_j = R_j$ for all j, which finishes the proof.

If we write (1) in the form

$$\sum_i a_i^2 (\tilde{M}_i - M_i) \le 0, \tag{12}$$

we deduce that $\inf\{\tilde{M}_i - M_i\} \le 0$, with equality only if $\tilde{R}_i = R_i$ for all i. Summing up the results we have shown the following

Theorem 23.1 (Strebel [5], [11]). *Let φ be a quadratic differential with closed trajectories on an arbitrary Riemann surface R. Assume that its trajectory structure determines finitely many punctured disks R_j, with punctures P_j, and possibly infinitely many ring domains R_k with finite moduli M_k. Moreover, let φ have finite reduced norm $\sum a_j^2 M_j + \sum_k a_k^2 M_k$, with M_j the reduced modulus of R_j. Let $\tilde{R}_j$, $\tilde{R}_k$ be a system of non overlapping punctured disks (with punctures P_j) and ring domains, of the homotopy type of the R_j and R_k respectively. Then, with M_j, $\tilde{M}_j$ the reduced moduli of the R_j, $\tilde{R}_j$ with respect to the same parameters near the second order poles P_j, the inequality*

$$\sum_j a_j^2 \tilde{M}_j + \sum_k a_k^2 \tilde{M}_k \le \sum_j a_j^2 M_j + \sum_k a_k^2 M_k \tag{I}$$

holds, with equality only if $\tilde{R}_j = R_j$, $\tilde{R}_k = R_k$ for all j and k. Moreover, as an easy consequence of (I)

$$\inf\{\tilde{M}_i - M_i\} \le 0, \tag{II}$$

with equality only if $\tilde{R}_i = R_i$ for all i.

23.2 Solution of the moduli problem for reduced moduli. The extremal property (II) can be used to prove the existence of certain quadratic differentials with second order poles. The simplest case is that of finitely many punctured disks R_j (and no additional ring domains R_k). Then, one derives from Theorem 21.7 that the reduced moduli M_j of the punctured disks R_j of φ can be prescribed arbitrarily, up to a common additive constant c. In other words, for arbitrarily given numbers m_j we can find φ such that

$$M_j = m_j + c \tag{1}$$

for all j, where the constant c is independent of j.

The reduced modulus M_j is related to the mapping radius r_j of R_j (with respect to the same parameter near the puncture P_j) by $M_j = \frac{1}{2\pi} \log r_j$. Therefore (1) goes over into

$$r_j = e^{2\pi c} e^{2\pi m_j} = t \cdot t_j, \tag{2}$$

with arbitrarily given positive numbers t_j. This shows that the ratio of the mapping radii of the R_j can be prescribed:

$$r_1 : r_2 : \ldots = t_1 : t_2 : \ldots. \tag{3}$$

The quadratic differential φ is then uniquely determined up to a positive constant factor (which of course does not change its trajectory structure and therefore the mapping radii of its punctured disks). We will prove the following theorem, the simples t case of which (triply punctured sphere) was treated by Teichmüller in [1].

Theorem 23.2 (Strebel [5], [11]). *Let $(\gamma_j)_{j=1,2,\ldots,p}$ be an admissible system of Jordan curves on a hyperbolic Riemann surface R, and let each γ_j bound a once punctured disk, with puncture P_j, which we may include into the surface. We fix local parameters z_j near the points P_j. Then, for arbitrarily given real numbers m_j, there is a quadratic differential φ with closed trajectories of homotopy type (γ_j), the characteristic ring domains R_j of which have reduced moduli (with respect to the given parameters z_j)*

$$M_j = m_j + c,$$

for some c independent of j. φ is uniquely determined up to a positive constant factor. In particular, the punctured disks R_j are uniquely determined.

Proof. To prove the uniqueness, let φ and $\tilde{\varphi}$ be two solutions, with $M_j = m_j + c$, $\tilde{M}_j = m_j + \tilde{c}$ for all j. As the system of punctured disks R_j belongs to a quadratic differential, we can apply (II) and get

$$\operatorname*{Min}_j \{\tilde{M}_j - M_j\} = \tilde{c} - c \leq 0. \tag{4}$$

Similarly, starting with $\tilde{\varphi}$, we get $c - \tilde{c} \leq 0$. Therefore $c = \tilde{c}$, hence $\tilde{R}_j = R_j$ for all j. But then, φ and $\tilde{\varphi}$ must have the same closed trajectories, which is only possible if $\tilde{\varphi}$ is a positive constant multiple of φ.

The existence is easily established by means of Theorem 21.7. It is evident that there exist disjoint punctured disks R'_j with reduced moduli $M'_j = m_j + c'$ for some c'. On the other hand, for any fixed j, $M'_j - m_j = c'$ is bounded above. By a normal family argument there exists a system (R_j) with maximal value c of all c'. We map the R_j conformally onto punctured disks $0 < |\zeta_j| < r_j$, with normalization $\frac{d\zeta_j}{dz_j}(0) = 1$ at the punctures P_j. Choose a number ρ such that $0 < \rho < r_j$ for all j and denote by $R_j(\rho)$ the ring domains corresponding to the annuli $\rho < |\zeta_j| < r_j$ on R. Let $M_j(\rho)$ be the modulus of $R_j(\rho)$. The system $(R_j(\rho))$ is a maximal system of cylinders with moduli vector in the direction $(M_1(\rho), \ldots, M_p(\rho))$ on the truncated surface $R(\rho)$. Otherwise we would have a system $(R'_j(\rho))$ with $M'_j(\rho) = (1+\varepsilon) M_j(\rho)$ for some $\varepsilon > 0$. Adding the disks $|\zeta_j| < \rho$ to the cylinders $R'_j(\rho)$ we would get a system of punctured disks R'_j with reduced moduli

$$M'_j \geq M'_j(\rho) + \frac{1}{2\pi} \log \rho > M_j(\rho) + \frac{1}{2\pi} \log \rho = M_j.$$

Therefore, the system $R_j(\rho)$ is associated with a quadratic differential φ_ρ. Its closed trajectories are the circles $|\zeta_j| = \text{const.}$, which are independent of ρ. But then, the φ_ρ are just the restrictions of a quadratic differential φ on R with the R_j as characteristic punctured disks.

23.3 The set $\Gamma(P_1, \ldots, P_p)$. Let R be a Riemann surface with a finite set of marked points P_j, $j = 1, \ldots, p$. Let $\dot{R} = R \setminus \{P_j\}$ be the surface R punctured at the points P_j. We denote by $\Gamma(P_1, \ldots, P_p)$ the set of holomorphic quadratic differentials φ with closed trajectories on $\dot{R}$ of homotopy type (γ_j), where γ_j is an arbitrarily small simple loop around P_j. We exclude the twice punctured sphere. The characteristic ring domains of $\varphi \in \Gamma$ are cylinders $R_j \subset \dot{R}$, where R_j has P_j as one of its boundary components; the other one is not point like. It is permitted that for some j there is no R_j (R_j is degenerate), and we also put $\varphi = 0 \in \Gamma$.

We introduce fixed parameters z_j in the neighborhoods of the points P_j, $z_j = 0 \leftrightarrow P_j$. The mapping radius of R_j with respect to the parameter z_j is denoted by r_j, its reduced modulus by $M_j = \frac{1}{2\pi} \log r_j$. If R_j is missing, we set $r_j = 0$ and consequently $M_j = -\infty$. We set, as before $a_j = \inf_{\{\gamma_j\}} \int_{\gamma_j} |\varphi(z)|^{\frac{1}{2}} |dz|$, where γ_j is an arbitrarily small simple loop around P_j. If R_j is not degenerate, a_j is equal to the φ-length of the closed trajectories around P_j; otherwise it is zero, because P_j is at most a first order pole of φ. The representation of φ in terms of z_j near P_j is

$$\varphi(z_j) = -\left(\frac{a_j}{2\pi}\right)^2 \frac{1}{z_j^2} + \ldots, \tag{1}$$

with $a_j > 0$ if and only if R_j is not degenerate.

Theorem 23.3. *The set $\Gamma(P_1, \ldots, P_p)$ is closed under locally uniform convergence on $\dot{R} = R \setminus \{P_j\}$. Moreover, the lengths a_j and mapping radii r_j are* *continuous. In other words, if* $\varphi_n \in \Gamma$, $(\varphi_n) \to \varphi$ *locally uniformly on* $\dot{R}$, *then* $\varphi \in \Gamma$, $a_{jn} \to a_j$ $r_{jn} \to r_j$ *and* $M_{jn} \to M_j$ *for every* j.

The sequences (a_{jn}) *are therefore necessarily bounded. Conversely, if this holds, the sequence* (φ_n) *contains a subsequence which converges locally uniformly on* $\dot{R}$.

Proof. Let $\varphi_n \in \Gamma$, $(\varphi_n) \to \varphi$ locally uniformly on $\dot{R}$. In the neighborhood U_j of P_j φ_n is represented by the Laurent series (1), i.e.

$$\varphi_n(z_j) = -\left(\frac{a_{jn}}{2\pi}\right)^2 \frac{1}{z_j^2} + \dots. \tag{2}$$

Therefore, $\varphi_n(z_j) \cdot z_j^2$ is holomorphic in U_j and, by the maximum principle, the sequence $\varphi_n(z_j) \cdot z_j^2$ converges uniformly on any compact subset of U_j. This shows, that the sequence (a_{jn}) converges and that φ has the representation (1), with $a_j = \lim_{n\to\infty} a_{jn} \geq 0$. By a similar argument we see that if the φ_n have at most a first order pole at P_j, so does φ.

Let us now consider the mapping radii r_{jn} of the R_{jn}. They are bounded by assumption. We therefore can find a subsequence, which we denote by (φ_n) again, such that $\lim_{n\to\infty} r_{jn} = \bar{r}_j$ exists, $j = 1, \dots, p$. Let $\bar{r}_j > 0$ and let g_{jn} be the mapping of the disk $|\zeta_{jn}| < r_{jn}$ onto R_{jn}, with $g_{jn}(0) = P_j$ and the normalization $\frac{dz_j}{d\zeta_{jn}}(0) = 1$ (the radius r_{jn} of the disk is then in fact the mapping radius of R_{jn} with respect to z_j). After passing to a subsequence again we can assume that the mappings g_{jn} converge, locally uniformly in $|\zeta_j| < \bar{r}_j$, to g_j: $|\zeta_j| < \bar{r}_j \to R_j$, $\frac{dz_j}{d\zeta_j}(0) = 1$. The representation of φ_n in terms of ζ_j is

$$\varphi_n(\zeta_j) = -\left(\frac{a_{jn}}{2\pi}\right)^2 \frac{1}{\zeta_{jn}^2} \left(\frac{d\zeta_{jn}}{d\zeta_j}\right)^2. \tag{3}$$

It follows, as in Section 21.2, that φ has the representation

$$\varphi(\zeta_j) = -\left(\frac{a_j}{2\pi}\right)^2 \frac{1}{\zeta_j^2}, \quad |\zeta_j| < \bar{r}_j. \tag{4}$$

We conclude that, if $\varphi \neq 0$, a_j cannot be zero unless $\bar{r}_j = 0$. Moreover, φ has closed trajectories of length a_j in R_j, and the mapping radius of R_j (with respect to z_j) is $r_j \geq \bar{r}_j$.

In order to show the converse inequality, let $\varphi \neq 0$ and $r_j > 0$. We introduce the natural parameter ζ_j of φ in R_j, with $\frac{d\zeta_j}{dz_j}(0) = 1$. Then

$$\varphi_n(\zeta_j) \to \varphi(\zeta_j) = -\left(\frac{a_j}{2\pi}\right)^2 \frac{1}{\zeta_j^2} \tag{5}$$

locally uniformly in $0 < |\zeta_j| < r_j$. We choose two radii r_j' and r_j'' with $0 < r_j' < r_j'' < r_j$. For all sufficiently large n, $\varphi_n(\zeta_j) \neq 0$ for $|\zeta_j| \leq r_j''$. On the other hand, φ_n has closed trajectories around P_j, in a sufficiently small neighborhood of P_j, as $a_{jn}>0$. We can then, because of the absence of zeroes, move a closed trajectory of φ_n outwards till it first meets $|\zeta_j|=r_j''$: let this be the trajectory α_{jn}. Because of the uniform convergence of φ_n to φ in $r_j'\leq|\zeta_j|\leq r_j''$ it is evident, from a consideration of the angles of intersection of α_{jn} with the closed trajectories $|\zeta_j|=\text{const.}$ of φ, that for sufficiently large n α_{jn} cannot meet $|\zeta_j|=r_j'$. Therefore $r_{jn}\geq r_j'$. This shows that $\lim\limits_{n\to\infty} r_{jn}\geq r_j$, hence $\bar{r}_j\geq r_j$.

Since we can start with an arbitrary subsequence of (φ_n), we have in fact shown that, with respect to locally uniform convergence $\varphi_n\to\varphi$, the lengths a_{jn} and, if $\varphi\neq 0$ (otherwise it does not make sense) the mapping radii r_{jn} are continuous, and we do no longer have to distinguish between $\bar{r}_j$ and r_j.

If $\varphi\neq 0$ and $r_j>0$, the characteristic ring domains R_{jn} tend to R_j, in the sense that the normalized mappings g_{jn} tend, locally uniformly, to g_j. In order to complete this result, we want to show that for $r_j=0$ $R_{jn}\to P_j$. Assume, first, that there is a disk D_j: $|z_j|\leq r$ contained in R_{jn} for every n. The normalized mapping $\zeta_{jn}=g_{jn}^{-1}(z_j)$, $g_{jn}(0)=0$, $\dfrac{d\zeta_{jn}}{dz_j}(0)=1$, maps D_j onto a domain containing zero and contained in $|\zeta_{jn}|<r_{jn}$. Because of Koebe's theorem, applied to D_j, this excludes $r_{jn}\to 0$. Therefore the boundary α_{jn} of R_{jn} has points in common with $|z_j|=r$, for all sufficiently large n. On the other hand we have seen that $a_{jn}\to 0$. Choose $r'>r$ such that $\varphi_n\to\varphi$ uniformly in $r\leq|z_j|\leq r'$. The two circles have a positive shortest distance in the φ-metric. This evidently leads to a contradiction if α_{jn} meets $|z_j|=r'$ for every n, showing that $R_{jn}\subset\{|z_j|<r'\}$ for all sufficiently large n. Since we can choose $0<r<r'$ arbitrarily, it follows that $R_{jn}\to P_j$.

In order to show that φ has closed trajectories, let $r_j>0$ for $j=1,\ldots,q$, while $r_j=0$ for $j=q+1,\ldots,p$. Clearly, $1\leq q\leq p$. We cut holes U_j: $|\zeta_j|<\rho$, $j=1,\ldots,q$ into the characteristic ring domains R_j of φ. Let $R_j^0=R_j\setminus U_j$, $M_j^0=\dfrac{1}{2\pi}\log\dfrac{r_j}{\rho}$, which is equal to the modulus of R_j^0. Because the trajectories of φ in U_j are closed, what we have to show is $\|\varphi\|_{R^0}=\sum\limits_{j=1}^{q}\|\varphi\|_{R_j^0}=\sum\limits_{j=1}^{q}a_j^2M_j^0$, where R^0 is the truncated surface $R^0=R\setminus\bigcup\limits_{j=1}^{q}U_j$. For given $\varepsilon>0$, let $r>0$ be so small that the neighborhoods U_j: $|z_j|<r$ of the points P_j, $j=q+1,\ldots,p$, cover a set of φ-area less than ε. Let $R^{00}=R\setminus\bigcup\limits_{j=1}^{p}U_j$. On R^{00}, the quadratic differentials φ_n are holomorphic, and the sequence (φ_n) converges locally uniformly to φ. Therefore $\|\varphi\|_{R^{00}}\leq\varliminf\limits_{n\to\infty}\|\varphi_n\|_{R^{00}}$. With $n\to\infty$, the ring domains $R_{jn}^0=R_{jn}\setminus\{|\zeta_{jn}|<\rho\}$, $j=1,\ldots,q$, cover R^{00} up to arbitrarily small annuli near the circles $|\zeta_j|=\rho$. Therefore, for sufficiently large n,

$$\|\varphi_n\|_{R^{00}}\leq\sum_{j=1}^{q}\|\varphi_n\|_{R_{jn}^0}+\varepsilon=\sum_{j=1}^{q}a_{jn}^2M_{jn}^0+\varepsilon, \tag{6}$$

with $M_{jn}^0=\dfrac{1}{2\pi}\log\dfrac{r_{jn}}{\rho}$ the modulus of R_{jn}^0. We end up with the inequalities

(7) $$\|\varphi\|_{R^0} \le \|\varphi\|_{R^{00}} + \varepsilon \le \lim_{n\to\infty} \|\varphi_n\|_{R^{00}} + \varepsilon$$

$$\le \overline{\lim_{n\to\infty}} \|\varphi_n\|_{R^{00}} + \varepsilon \le \lim_{n\to\infty} \sum_{j=1}^{q} a_{jn}^2 M_{jn}^0 + 2\varepsilon$$

$$\le \sum_{j=1}^{q} a_j^2 M_j^0 + 2\varepsilon \le \|\varphi\|_{R^0} + 2\varepsilon.$$

We thus have the equality

(8) $$\|\varphi\|_{R^0} = \sum_{j=1}^{q} a_j^2 M_j^0 = \sum_{j=1}^{q} \|\varphi\|_{R_j^0},$$

which shows that the ring domains R_j^0 cover R^0 up to a set of φ-area zero, and hence of measure zero. But then $R \setminus \bigcup_{j=1}^{q} R_j$ is a null set too.

To show the last statement of the theorem, let $\varphi_n \in \Gamma(P_1, \ldots, P_p)$ for $n = 1, 2, \ldots$ and let the sequences $(a_{jn})_{n=1,2,\ldots}$ be bounded for all $j = 1, \ldots, p$. Let U_j: $|z_j| < r$, $j = 1, \ldots, p$, and let $R^0 = R \setminus \bigcup_{j=1}^{p} U_j$. A simple normal family argument shows that for each fixed j there is a number m_j with the following property: any ring domain $\tilde{R}_j$ of homotopy type γ_j with modulus $\tilde{M}_j > m_j$ has at least one of its boundary components lying in U_j. For every characteristic ring domain R_{jn} of φ_n we now choose $\rho_{jn} > 0$ such that

(9) $$m_j < \frac{1}{2\pi} \log \frac{r_{jn}}{\rho_{jn}} < 2m_j.$$

The subring R_{jn}^0 of R_{jn} which corresponds to the annulus $\rho_{jn} < |\zeta_{jn}| < r_{jn}$ then has at least one of its boundary components in U_j. Therefore the ring domains R_{jn}^0 cover R^0 up to a set of measure zero, and consequently

(10) $$\|\varphi_n\|_{R^0} \le \sum_j \|\varphi_n\|_{R_{jn}^0} \le 2 \sum_j a_{jn}^2 m_j,$$

which is bounded independently of n. We conclude, that the sequence (φ_n) is locally bounded in $\dot{R}$ and therefore has a subsequence which converges locally uniformly on $\dot{R}$ to a quadratic differential φ. This finishes the proof of the theorem.

23.4 The surface of reduced moduli. In this section we consider the subset of those $\varphi \in \Gamma(P_1, \ldots, P_p)$ for which all the mapping radii r_j are positive. We denote by

$$x = (x_1, \ldots, x_p) = (M_1, \ldots, M_p)$$

the p-tupel of reduced moduli M_j of R_j (with respect to some fixed parameters z_j near P_j), and by

$$y = (y_1, \ldots, y_p) = (a_1^2, \ldots, a_p^2)$$

the p-tupel of the squares of the lengths a_j of the closed trajectories of φ. We are interested in the surface $\mathcal{M}$ described by the vector x, $p \ge 2$.

Let $\bar{M}_j$ be the maximal reduced modulus of all ring domains on $\dot{R}$ with point boundary component P_j. The surface $\mathcal{M}$ lies in the intersection of the half spaces $x_j < \bar{M}_j$. For its description, we use $m = (m_1, \ldots, m_p)$ as parameter, normalised such that $\sum_{j=1}^{p} m_j = 0$. By the existence theorem of Section 23.2 there is associated, with every m of the above plane, a unique vector x, $x_j = m_j + c$. This means that the surface $\mathcal{M}$ cuts every normal to the plane $\sum m_j = 0$ in exactly one point (Fig. 65).

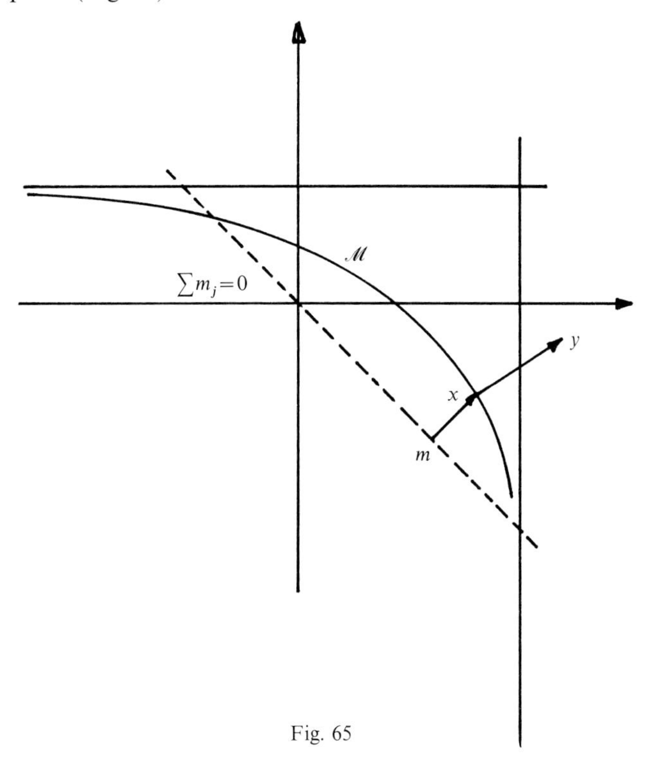

Fig. 65

We first show the continuity of x as a function of m. To this end, we pass to the corresponding quadratic differentials $\varphi \in \Gamma(P_1, \ldots, P_p)$, normalized (by multiplication with a positive constant) such that $\|y\| = \sqrt{\sum a_j^4} = 1$.

Let now $m^{(n)} = (m_{1n}, \ldots, m_{pn}) \to m$. Let $\varphi_n \in \Gamma$ be the associated normalized quadratic differential (with reduced moduli $M_{jn} = m_{jn} + c_n$). Since its lengths a_{jn} are bounded by the requirement that $\|y^{(n)}\| = \sqrt{\sum a_{jn}^4} = 1$, there exists a subsequence which converges locally uniformly on $\dot{R}$ to a quadratic differential $\varphi \in \Gamma$. By the continuity of the reduced moduli, φ has the moduli vector $M_j = m_j + c$; it is therefore the unique solution of the moduli problem for m. Since we can start with any subsequence of (φ_n), we have in fact shown that $M^{(n)} \to M$, i.e. $x^{(n)} \to x$.

To show the convexity of $\mathcal{M}$ and the fact that y is the (continuous) unit normal, we proceed as in Sections 21.6 and 21.9. For an arbitrary $x \in \mathcal{M}$, let

$\tilde{x} \in \mathcal{M}$, $\tilde{x} \neq x$. Then, by Theorem 23.1, $(\tilde{x}, y) < (x, y)$. Therefore $(\tilde{x} - x, y) < 0$, and with the notation $e = \frac{\tilde{x} - x}{\|\tilde{x} - x\|}$, $(e, y) < 0$. On the other hand, because of the continuity of the lengths a_j, the unit vector y is continuous. We thus have $\tilde{y} = y + \varepsilon$ with $\varepsilon \to 0$ for $\tilde{x} \to x$. From $(x, \tilde{y}) < (\tilde{x}, \tilde{y})$ we then get $(\tilde{x} - x, \tilde{y}) > 0$, hence $(e, \tilde{y}) = (e, y + \varepsilon) > 0$, and finally $(e, y) > -(e, \varepsilon)$. This shows that the surface $\mathcal{M}$ is strictly convex and that the vector $y = (a_1^2, \ldots, a_p^2)$ is its normal. We sum it up in the following

Theorem 23.4. *Let R be a Riemann surface with distinguished points P_j, $j = 1, \ldots, p$, $p \geq 2$ and $\dot{R} = R \setminus \{P_j\}$ not the twice punctured sphere. We consider the subset of quadratic differentials $\varphi \in \Gamma(P_1, \ldots, P_p)$ with non vanishing characteristic ring domains R_j around the points P_j. Then, for every point $m = (m_1, \ldots, m_p)$ of the plane $\sum_{j=1}^{p} m_j = 0$ there is exactly one p-tupel of reduced moduli $M = (M_1, \ldots, M_p)$ of the form $M_j = m_j + c$, $j = 1, \ldots, p$. M is a continuous function of m and the surface $\mathcal{M}$ so described is strictly convex and has a continuous normal unit vector $y = (a_1^2, \ldots, a_p^2)$, where $a_j > 0$ is the length of the trajectories of φ around P_j. (φ is normalized by a positive factor such that $\|y\| = 1$.)*

23.5 Quadratic differentials with given lengths of the trajectories. The surface $\mathcal{M}$ of reduced moduli allows the solution of the following problem: given the lengths $A_j > 0$ of the closed trajectories around the points P_j, find the quadratic differential $\varphi \in \Gamma(P_1, \ldots, P_p)$ with $a_j = A_j$. Equivalently: given the negative leading coefficients of the second order poles P_j. Find the corresponding quadratic differential φ with closed trajectories.

The procedure is the following: One pushes the plane with normal $Y = (A_1^2, \ldots, A_p^2)$ from the right to the surface $\mathcal{M}$. It touches $\mathcal{M}$ at a uniquely determined point $M = (M_1, \ldots, M_p)$. There is a quadratic differential $\varphi \in \Gamma$ with reduced moduli M_j. It is uniquely determined up to a positive factor, and the vector $y = (a_1^2, \ldots, a_p^2)$ is normal to $\mathcal{M}$ at M and hence parallel to Y. Therefore $A_j^2 = \lambda \cdot a_j^2$, $j = 1, \ldots, p$. The trajectories of $\lambda \cdot \varphi$ have lengths $\sqrt{\lambda} \cdot a_j = A_j$, as prescribed.

Theorem 23.5. *Given a Riemann surface R with marked points P_j, $j = 1, \ldots, p$, $p \geq 2$ and $\dot{R} = R \setminus \{P_j\}$ not the twice punctured sphere. We consider the quadratic differentials φ on $\dot{R}$ with closed trajectories the characteristic ring domains of which are punctured disks R_j, with punctures P_j. Then, the lengths $a_j > 0$ of the closed trajectories α_j around the P_j can be prescribed arbitrarily. The solution φ is uniquely determined. This amounts to the same as prescribing the negative leading coefficients of the Laurent series of φ at the points P_j.*

Chapter VII. Quadratic Differentials of General Type

§24. An Extremal Property for Arbitrary Quadratic Differentials

24.1 The height of a loop or a cross cut. For any rectifiable closed curve γ on a Riemann surface R and holomorphic quadratic differential φ on R the φ-length

$$|\gamma|_{\varphi}=\int_{\gamma}|\varphi(z)|^{\frac{1}{2}}\,|dz|$$

is defined. Rectifiability is meant with respect to the local parameters z on R; it is evidently independent of the choice of the local parameter. The infimum of the φ-lengths for all loops $\tilde{\gamma}$ in the free homotopy class of γ is denoted by

$$\ell_{\varphi}(\gamma)=\inf_{\tilde{\gamma}\sim\gamma}\int_{\tilde{\gamma}}|\varphi(z)|^{\frac{1}{2}}\,|dz|.$$

It turns out to be important to consider not only the quantities $|\gamma|_{\varphi}$ and $\ell_{\varphi}(\gamma)$, but also the integrals over the absolute values of the real and imaginary part of $\varphi(z)^{\frac{1}{2}}dz$, i.e. the integrals $\int_{\gamma}|\mathrm{Re}\{\varphi(z)^{\frac{1}{2}}dz\}|$ and $\int_{\gamma}|\mathrm{Im}\{\varphi(z)^{\frac{1}{2}}dz\}|$. To fix the ideas, we stick to the latter. Because of the continuity of φ the integral exists, but we evidently only have to consider the subintervals of γ between the zeroes of φ. At the corresponding points of R the function $w=u+iv=\Phi(z)$ $=\int_{\gamma}\sqrt{\varphi(z)}\,dz$ is defined, and we can thus write

$$\int_{\gamma}|\mathrm{Im}\,\{\varphi(z)^{\frac{1}{2}}\,dz\}|=\int_{\gamma}|dv|$$

which is the total variation of the imaginary part v of Φ.

Definition 24.1. Let φ be a holomorphic quadratic differential on an arbitrary Riemann surface R. Then, for any closed curve γ, the infimum

$$h_{\varphi}(\gamma)=\inf_{\tilde{\gamma}\sim\gamma}\int_{\tilde{\gamma}}|dv|,\qquad w=\Phi(z)=u+iv, \tag{1}$$

where $\tilde{\gamma}$ varies over all rectifiable closed curves of the free homotopy class of γ, is called the φ-height of γ, or simply its height.

In general, there is no curve $\gamma_0\sim\gamma$ for which we have equality

$$h_{\varphi}(\gamma)=\int_{\gamma_0}|dv|. \tag{2}$$

A simple counterexample is an annulus R with $\varphi \equiv 1$ and γ a loop separating the two boundary components of R. But if the free homotopy class of γ contains curves γ_0 of a special kind, then (2) is realized.

Theorem 24.1. *Let $\gamma_0 \sim \gamma$ be a step curve, i.e. a geodesic polygon the sides of which are horizontal and vertical arcs of φ, with the additional property that for any horizontal side α_1 of γ the two neighboring vertical sides β_1, and β_2 are on different sides of α_1 (there are no zeroes of φ on γ). Then*

$$h_\varphi(\gamma) = \int_{\gamma_0} |dv|.$$

The same is true if γ_0 is a geodesic in the free homotopy class of γ. On the other hand, for every γ and $\varepsilon > 0$ there is a step curve $\gamma_0 \sim \gamma$ which does not go through a zero of φ and such that

$$\int_{\gamma_0} |dv| < h_\varphi(\gamma) + \varepsilon.$$

Proof. Let γ_0 be a step curve in the free homotopy class of γ, satisfying the additional condition of the theorem (Fig. 66). Every horizontal side α_i of γ has

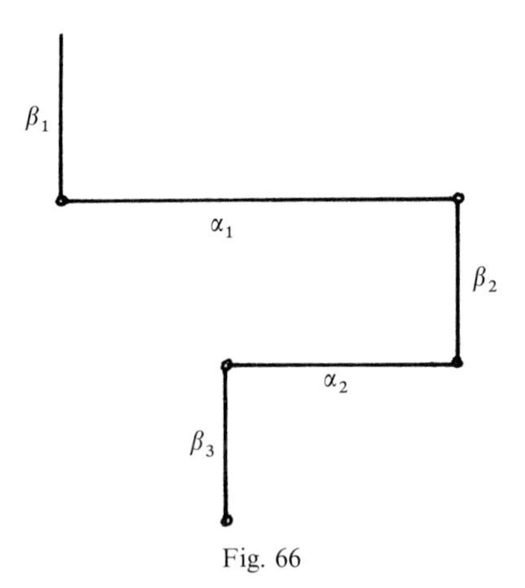

Fig. 66

two vertical sides β_i and β_{i+1}, say, as neighbors. The interior angles at the two vertices are $\pi/2$ and $\frac{3\pi}{2}$ respectively. Teichmüller's lemma for a simple closed geodesic polygone and a holomorphic φ reads

$$\sum_j \left(1 - \vartheta_j \frac{n_j + 2}{2\pi}\right) \geq 2. \tag{3}$$

Since there are no zeroes on γ_0, all the n_j are zero, hence $\sum_j \left(1 - \frac{1}{\pi} \vartheta_j\right) \geq 2$. The contributions from the two vertices at the endpoints of α_i are $(1 - \frac{1}{2}) + (1 - \frac{3}{2}) = 0$. Therefore, γ_0 cannot bound a simply connected domain on R.

Let $\hat{R}(\gamma)$ be the annular covering surface associated with γ. The lifts $\hat{\gamma}$ and $\hat{\gamma}_0$ respectively have winding number one around the interior boundary component of $\hat{R}(\gamma)$. We claim that $\hat{\gamma}_0$ is simple. Otherwise $\hat{\gamma}_0$ contains a subinterval which forms a simple loop bounding a simply connected subdomain of $\hat{R}(\gamma)$. Let z be the point where it closes (intersection point of $\hat{\gamma}_0$ with itself). By a small parallel shift of one or both of thc sides meeting at z we can achieve that z is an interior point of a horizontal and a vertical arc. Then, the angle $\frac{3\pi}{2}$ can be replaced by $\frac{\pi}{2}$ which gives a contribution one instead of zero at the left hand side of (3). There still is a contradiction. Thus $\hat{\gamma}_0$ is a simple loop separating the two boundary components of $\hat{R}(\gamma)$.

Let $\hat{\alpha}$ be a regular trajectory of the lift $\hat{\varphi}$ of φ in $\hat{R}(\gamma)$, and assume that it has a point z_1 in common with $\hat{\gamma}_0$. $\hat{\alpha}$ is a cross cut of $\hat{R}(\gamma)$. Assume that both of its ends tend to the same boundary component of $\hat{R}(\gamma)$. Then it has another point z_2 in common with γ_0 and, by considering a neighboring trajectory of $\hat{\alpha}$, if necessary, we may assume that z_1 and z_2 are interior points of vertical sides of $\hat{\gamma}_0$. A subinterval of $\hat{\gamma}_0$ together with a subinterval of $\hat{\alpha}$ then forms a step curve, bounding a simply connected subdomain of $\hat{R}(\gamma)$. We have the same situation as above, namely possibly two interior angles $\frac{\pi}{2}$ at the endpoints of the subinterval of $\hat{\alpha}$, which contradicts Teichmüller's lemma. We conclude that every regular horizontal trajectory $\hat{\alpha}$ of $\hat{\varphi}$ which meets $\hat{\gamma}_0$ is "radial", i.e. connects the two boundary components of $\hat{R}(\gamma)$.

Let γ_1 be a rectifiable closed curve on R freely homotopic to γ. Its lift $\hat{\gamma}_1$ on $\hat{R}(\gamma)$ separates the two boundary components of $\hat{R}(\gamma)$. We choose two radii $r_1 < r_2$ such that the annulus $\hat{R}_1$: $r_1 \leq |z| \leq r_2$ is contained in $\hat{R}(\gamma)$ and contains both curves $\hat{\gamma}_0$ and $\hat{\gamma}_1$. There are only finitely many zeroes of $\hat{\varphi}$ in $\hat{R}_1$ and therefore only finitely many horizontal trajectories through points of $\hat{\gamma}_0$ which are critical in $\hat{R}_1$. We thus have finitely many open "radial" horizontal strips $\hat{S}_j$ in $\hat{R}_1$ cutting $\hat{\gamma}_0$. $\hat{S}_j$ has a vertical interval β_j in common with $\hat{\gamma}_0$, and the sum of the heights of the $\hat{S}_j$ is evidently equal to the sum of the lengths of the vertical intervals β_i of $\hat{\gamma}_0$, which is equal to $\int_{\hat{\gamma}_0} |d\hat{v}|$. On the other hand, every $\hat{S}_j$ is traversed by $\hat{\gamma}_1$, and of course the $\hat{S}_j$ are non overlapping. Therefore

$$\int_{\gamma_1} |dv| = \int_{\hat{\gamma}_1} |d\hat{v}| \geq \int_{\hat{\gamma}_0} |d\hat{v}| = \int_{\gamma_0} |dv|. \tag{4}$$

This proves that equality (2) holds.

The proof for a geodesic loop $\gamma_0 \sim \gamma$ is essentially the same. The lift $\hat{\gamma}_0$ of γ_0 to $\hat{R}(\gamma)$ is simple (because of the uniqueness of geodesic connections), and, by the same reason, any trajectory $\hat{\alpha}$ which does not lie on $\hat{\gamma}_0$ has at most one point in common with $\hat{\gamma}_0$. We can therefore argue in the same way as above: Every side of $\hat{\gamma}_0$ which is not horizontal is covered (up to finitely many points) by a certain number of open horizontal strips $\hat{S}_j$. The total height of these strips is equal to the integral $\int |dv|$ over the corresponding side of $\hat{\gamma}_0$. On the other hand, every strip is crossed by the lift $\hat{\gamma}_1$ of γ_1. We again deduce the inequality (4), hence equality (2).

To show the last statement of the theorem, we choose, for given $\varepsilon > 0$, a rectifiable loop $\gamma_1 \sim \gamma$ such that

(5) $$\int_\gamma |dv| < h_\varphi(\gamma) + \varepsilon.$$

By a slight change we can push γ_1 away from the possible finitely many zeroes on it, without vexing (5). We now subdivide it into finitely many subintervals, the endpoints of which are connected by pairs of vertical and horizontal sides (one of which may degenerate to a point). The new curve γ_0 is a step curve with $\int_{\gamma_0} |dv| \le \int_{\gamma_1} |dv|$, and γ_0 does not contain any zero of φ if the subdivision of γ_1 is small enough. Of course γ_0 will in general not satisfy the additional condition of the theorem.

As is mentioned in the title of the section, the definition of height can be extended to homotopy classes of cross cuts. The following situation is considered later in the chapter: Let R be a Riemann surface with a pair of distinguished open boundary arcs γ_1 and γ_2 (being part of the surface), along which $\varphi(z)\,dz^2$ is real. Let γ be a rectifiable curve on R joining γ_1 and γ_2. A continuous deformation of γ is meant to move each endpoint of it on the respective boundary arc. The height $h_\varphi(\gamma)$ is then defined in the same way. By welding R with its reflection on $\gamma_1 \cup \gamma_2$, the curve γ becomes a closed curve, and this situation can thus be reduced to the former one.

It is also possible to consider curves γ with fixed endpoints, but we will not deal with this situation.

24.2 The extremal property on compact surfaces with punctures. For a holomorphic quadratic differential φ on an arbitrary Riemann surface R the function $\Phi(z) = \int \sqrt{\varphi(z)}\,dz$ is locally defined, outside of the zeroes of φ, up to its sign and an arbitrary additive constant. If we set $w = u + iv = \Phi(z)$, we have local function elements v, with the relation

(1) $$v_2 = \pm v_1 + \text{const.}$$

for two elements v_1 and v_2 in the same neighborhood. We also find $\inf_{\tilde\gamma \sim \gamma} \int_{\tilde\gamma} |dv| = h_\varphi(\gamma)$ and

(2) $$\|dv\|^2 = D(v) \equiv \iint_R \left(\left(\frac{\partial v}{\partial x}\right)^2 + \left(\frac{\partial v}{\partial y}\right)^2 \right) dx\,dy = \|\varphi\|,$$

i.e. the Dirichlet integral over v is equal to the norm of φ.

Let now $\{\tilde v\}$ be a set of local C^1-function elements on R, defined outside of an exceptional set E with no accumulation points in R, and such that two different function elements $\tilde v_1$ and $\tilde v_2$ in the same neighborhood on R satisfy (1). The Dirichlet integral $D(\tilde v)$ has an invariant meaning, and we can again set $\|d\tilde v\|^2 = D(\tilde v)$. Moreover, the notion of height of a closed curve or cross cut γ carries over to this situation. We define

$$h_{\tilde v}(\gamma) = \inf_{\tilde\gamma \sim \gamma} \int_{\tilde\gamma} |d\tilde v|,$$

where $\tilde\gamma$ is an arbitrary rectifiable loop or locally rectifiable cross cut in the free homotopy class of γ.

Theorem 24.2. *Let R be a compact Riemann surface and let φ be a meromorphic quadratic differential on R with finite norm. The surface R punctured at the poles P_j of φ (necessarily of the first order) is denoted by $\dot{R}=R\setminus\{P_j\}$. Let $\{\tilde{v}\}$ be a set of C^1-function elements on $\dot{R}\setminus$(exceptional set) with the invariance property* (1) *and satisfying the height condition*

$$h_{\tilde{v}}(\gamma)\geq h_{\varphi}(\gamma) \tag{3}$$

for every simple closed loop γ on $\dot{R}$. (The homotopy is of course also meant on the punctured surface R.) Then, the norm inequality

$$\|d\tilde{v}\|^2=D(\tilde{v})\geq\|\varphi\| \tag{4}$$

holds, with equality only for $\{\tilde{v}\}=\{v\}$.

Proof. The proof is based on the vertical trajectory structure of φ. (If $\varphi=0$, there is nothing to prove.) We first consider a ring domain R_i swept out by vertical trajectories, cut it along a horizontal cross cut and map the quadrilateral by means of $w=u+iv=\Phi(z)$ onto a vertical rectangle S_i with sides a_i, b_i. We then express the function elements $\tilde{v}(u,v)$ in terms of the variables u, v outside of the exceptional set in S_i. For $\gamma=\beta_i$ a closed vertical trajectory in R_i we have, by Theorem 24.1, $h_{\varphi}(\gamma)=\int_{\beta_i} dv=b_i$. Therefore, by assumption,

$$b_i\leq\int_{\beta_i}|d\tilde{v}|=\int_0^{b_i}\left|\frac{\partial\tilde{v}}{\partial v}\right|dv \tag{5}$$

for every vertical β_i in S_i which does not meet the exceptional set for $\tilde{v}$. Integration over u yields

$$\|\varphi\|_{R_i}=a_i b_i\leq\iint_{S_i}\left|\frac{\partial\tilde{v}}{\partial v}\right|du\,dv. \tag{6}$$

Second, we consider a vertical spiral set B_i, i.e. the closure of a recurrent vertical trajectory. A horizontal interval α in B_i induces a partitioning of B_i into vertical strips S_ν of the first kind (Fig. 67) and pairs of vertical strips S_μ^+ and S_μ^- of equal width of the second kind (Fig. 68).

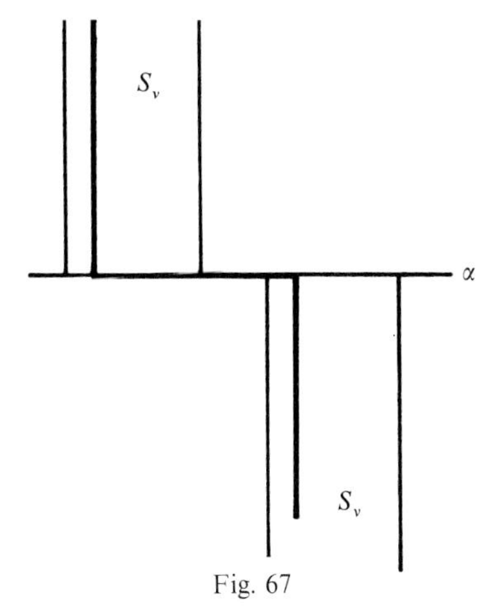

Fig. 67

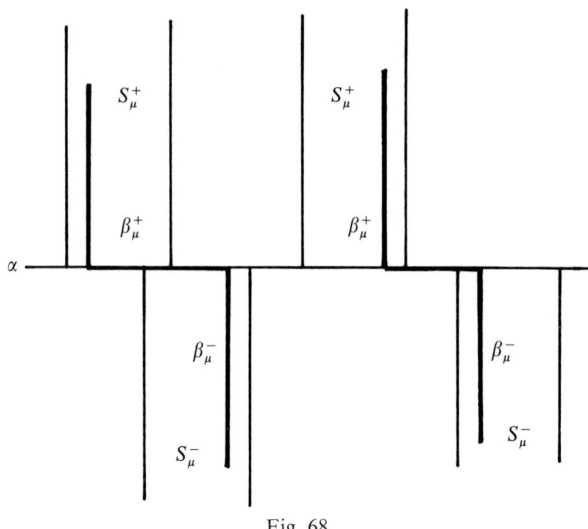

Fig. 68

A vertical arc β_ν in S_ν can be transformed, by means of a horizontal connection of its endpoints, into a simple loop γ. Its φ-height is, by Theorem 24.1, equal to $h_\varphi(\gamma)=\int_{\beta_\nu} dv=b_\nu$. In the second case, two vertical arcs β_μ^+ and β_μ^- in S_μ^+ and S_μ^- can again be tied together, along α, to form a loop γ. The connecting intervals on α have to be chosen in such a way that a small lift of one of them transforms γ into a simple loop. Because the additional condition of Theorem 24.1 is satisfied, we again have $h_\varphi(\gamma)=b_\mu^+ + b_\mu^-$, the b's denoting the φ-lengths of the vertical arcs β_μ^+ and β_μ^- respectively.

We now choose a neighborhood U in the interior of B_i such that we have function elements Φ and $\tilde{v}$ in U. Let M be a bound for $\left|\frac{\partial \tilde{v}}{\partial u}\right|$ in U. For given $\varepsilon>0$, let α be a horizontal interval in U with φ-length $|\alpha|_\varphi=a<\varepsilon$. Then, in S_ν and outside of the exceptional set for $\tilde{v}$ we have, because of $\int_\alpha \left|\frac{\partial \tilde{v}}{\partial u}\right| du \leq M\cdot a$,

$$b_\nu \leq \int_{\beta_\nu} \left|\frac{\partial \tilde{v}}{\partial v}\right| dv+\varepsilon\cdot M. \tag{7}$$

Integration over u yields

$$a_\nu b_\nu \leq \iint_{S_\nu} \left|\frac{\partial \tilde{v}}{\partial v}\right| du\, dv+a_\nu\cdot \varepsilon M, \tag{8}$$

where a_ν is the φ-length of the horizontal side of the rectangle S_ν.

Similarly we proceed for a pair of strips of the second kind. We get

$$b_\mu^+ + b_\mu^- \leq \int_{\beta_\mu^+} \left|\frac{\partial \tilde{v}}{\partial v}\right| dv+\int_{\beta_\mu^-} \left|\frac{\partial \tilde{v}}{\partial v}\right| dv+2\varepsilon M. \tag{9}$$

This is true for any β_μ^+ and β_μ^- in S_μ^+ and S_μ^- respectively, and because of the equal width a_μ of both rectangles we can integrate over u. This gives

(10)
$$(b_\mu^+ + b_\mu^-)\, a_\mu \le \iint_{S_\mu^+ \cup S_\mu^-} \left|\frac{\partial \tilde{v}}{\partial v}\right| du\, dv + a_\mu \cdot 2\varepsilon M.$$

Summing over all the strips in B_i we get

(11)
$$\|\varphi\|_{B_i} \le \iint_{B_i} \left|\frac{\partial \tilde{v}}{\partial v}\right| du\, dv + a \cdot \varepsilon \cdot M.$$

It is important that the integral on the right hand side, although not invariant under an arbitrary change of the parameter on R, has an invariant meaning with respect to the parameter $w = u + iv$ and an arbitrary subdivision of B_i into patches with single valued Φ. Therefore α can be made arbitrarily small within U, and we get

(12)
$$\|\varphi\|_{B_i} \le \iint_{B_i} \left|\frac{\partial \tilde{v}}{\partial v}\right| du\, dv.$$

Summing up (6) and (12) over all the finitely many ring domains R_i and spiral sets B_i on R we get

(13)
$$\|\varphi\| \le \iint_R \left|\frac{\partial \tilde{v}}{\partial v}\right| du\, dv.$$

We now apply Schwarz's inequality and add a term in the integrand. We get, successively,

(14)
$$\begin{aligned}\|\varphi\|^2 &\le \|\varphi\| \cdot \iint_R \left(\frac{\partial \tilde{v}}{\partial v}\right)^2 du\, dv \\ &\le \|\varphi\| \iint_R \left(\left(\frac{\partial \tilde{v}}{\partial u}\right)^2 + \left(\frac{\partial \tilde{v}}{\partial v}\right)^2\right) du\, dv.\end{aligned}$$

This proves the inequality (4). Let equality hold. Then, $\dfrac{\partial \tilde{v}}{\partial u} \equiv 0$ on $\dot{R} \setminus E$, where E is the exceptional set for $\tilde{v}$. Moreover, $\dfrac{\partial \tilde{v}}{\partial v}$ must be constant on $\dot{R} \setminus E$, and the constant clearly must be one. Because of continuity we get, in every single neighborhood with well defined Φ and $\tilde{v}$, $\dfrac{\partial \tilde{v}}{\partial v} \equiv 1$ or $\equiv -1$, hence

(15)
$$\tilde{v} = \pm v + \text{const.}$$

The function elements $\tilde{v}$ can then be extended to $R \setminus \{\text{zeroes of } \varphi\}$, therefore the two sets of function elements $\{\tilde{v}\}$ and $\{v\}$ are the same. This finishes the proof of the theorem.

It should be noticed that one can start with a compact Riemann surface R and given punctures P_j. We consider a quadratic differential φ which is holomorphic on $\dot{R} = R \setminus \{P_j\}$ and has finite norm. The punctures P_j which are regular points of φ then can be neglected, i.e. some of the free homotopy classes on R coincide.

24.3 The extremal property on compact bordered surfaces with punctures. The theorem of the last section can be generalized to the case of a compact bordered surface and a quadratic differential which is real along the boundary

∂R of R. φ can have first order poles and zeroes on the boundary curves, but in the intervals between we have $\varphi(z)\,dz^2>0$ or $\varphi(z)\,dz^2<0$ for tangential dz.

The trajectory structure is most easily recognized if one passes to the double $\tilde{R}$ of R. Since φ is real along ∂R, it can be continued to $\tilde{R}$ by reflection. The result is a symmetric surface $\tilde{R}$ with a symmetric quadratic differential $\tilde{\varphi}$. Also its vertical trajectory structure is symmetric. Let β^+ be a regular vertical trajectory ray with initial point P_1 on a horizontal boundary interval, which points into R. It evidently cannot stay in R. Let P_2 be its first intersection point with ∂R after P_1. Then the subarc P_1, P_2 of β forms, together with its mirror image, a closed vertical trajectory. Therefore the structure of φ on R consists of vertical ring domains, vertical spiral sets and vertical rectangles with both horizontal sides on ∂R. The curves to be considered are therefore, besides simple closed loops on the interior of R, punctured at the points P_j, cross cuts connecting open horizontal boundary intervals. The proof of the last section then goes through, with the vertical rectangles as additional subdomains to be considered. We have

Theorem 24.3. *Let R be a compact bordered surface and let φ be a meromorphic quadratic differential on R with finite norm, which is real along ∂R. Let $h_{\tilde{v}}(\gamma)\geq h_{\varphi}(\gamma)$ for every simple loop γ on $\dot{R}$ and every cross cut which connects two open horizontal intervals on ∂R. Then $\|\partial\tilde{v}\|^2\geq\|\varphi\|$, with equality only if the two sets of function elements $\{\tilde{v}\}$ and $\{v\}$ are the same.*

24.4 Parabolic surfaces. The norm inequality (4) stated in Theorem 24.2 can be generalized to the case where the vertical cross cuts of φ (i.e. the vertical trajectories both ends of which tend to the boundary of the surface) cover a set of measure zero only. It is therefore of interest to know the class of surfaces for which this is true for every holomorphic quadratic differential $\varphi\neq0$ of finite norm. In this section we only give a sufficient condition for this property of R and formulate it for horizontal trajectories.

Theorem 24.4. *Let R be a parabolic Riemann surface (i.e. R has no Green's function). Then every holomorphic quadratic differential $\varphi\neq0$ of finite norm has the property that the horizontal trajectories which are cross cuts of R cover a set of measure zero.*

Proof. Let R be parabolic and let φ be a holomorphic quadratic differential in R, not identically equal to zero, and of finite norm. We choose an arbitrary closed, oriented vertical interval β. Since R is parabolic, the module of the family of curves joining β to the boundary ∂R of R is zero (see e.g. Letho and Virtanen [1] and Strebel [1]). Let $E\subset\beta$ be the set of initial points of boundary rays α^+ of φ leaving β to the right and not cutting β again. They are mapped by a branch of Φ onto a set S' of half open horizontal intervals in the $w=u+iv$-plane leaving the vertical interval $\beta'=\Phi(\beta)$ to the right. The inverse Φ^{-1} is a $1-1$ conformal mapping of S' onto the set of rays α^+. Since this family $\{\alpha^+\}$ has module zero, so does the family of horizontal intervals in S'. We denote by $E'=\Phi(E)$ the set of their initial points on β'.

Let $\rho \geq 0$ be a measurable metric on S' with the property that $\int \rho\, du \geq 1$ on every horizontal $\Phi(\alpha^+)$ in S'. We get, by integrating over E',

(1) $$|E'| \leq \iint_{S'} \rho\, du\, dv$$

and by Schwarz's inequality

(2) $$|E'|^2 \leq |S'| \cdot \iint_{S'} \rho^2\, du\, dv.$$

Here, the bars denote one and two dimensional Lebesgue measure respectively. Since (2) is true for every ρ and the module of the curve family $\{\alpha^+\}$ is equal to

(3) $$m\{\alpha^+\} = \inf \iint_{S'} \rho^2\, du\, dv = 0,$$

$|S'| = \|\varphi\|_S \leq \|\varphi\| < \infty$, we conclude that $|E'| = 0$. Therefore S' and hence S has two dimensional measure zero.

We now choose a denumerable set of regular points P which is dense on R, and through each point P a denumerable set of vertical intervals β of arbitrarily small length. The totality of half strips $S(\beta)$ covers a set of measure zero on R. Let P_1 be a point on a horizontal cross cut α. P_1 is the center of a horizontal square Q such that $\alpha \cap Q$ is an interval containing P_1 as its midpoint. Evidently P and β through P can be chosen in Q such that the half strip $S'(\beta)$ contains P_1. Therefore the set of horizontal cross cuts covers a set of measure zero.

24.5 The extremal property on parabolic surfaces. The study of the trajectory structure on an arbitrary Riemann surface as exposed in §13, in particular Section 13.6, allows a generalization of Theorem 24.2 to open surfaces. The crucial property is that the vertical cross cuts, i.e. the vertical trajectories which tend to the ideal boundary of the surface in both of their directions, cover a set of measure zero only.

Theorem 24.5. *Let φ be a holomorphic quadratic differential of finite norm on an arbitrary Riemann surface R. Assume that the vertical cross cuts of φ cover a set of measure zero only. Let $\{\tilde{v}\}$ be as in Section 24.2, with*

(1) $$h_{\tilde{v}}(\gamma) \geq h_\varphi(\gamma)$$

for every simple closed loop γ on R. Then

(2) $$\|d\tilde{v}\|^2 \geq \|\varphi\|$$

holds, with equality only if $\{\tilde{v}\} = \{v\}$.

The conclusion (2) also holds on a bordered surface R if $\varphi\, dz^2$ is real along the border. In this case (1) must also be satisfied for the arcs joining boundary intervals of R.

Proof. For the (possibly denumerably many) ring domains swept out by closed vertical trajectories of φ there is no change. For the set of R covered by vertical spirals (Section 13.1c) we proceed as follows: let $\varepsilon > 0$ and choose a compact set $C \subset R$ which does not contain any exceptional points of the system

$\{\tilde{v}\}$ or critical points of φ and such that $\|\varphi\|_C > \|\varphi\| - \varepsilon$. Let $O \subset R$ be an open set containing C but no exceptional or critical points either and such that there exists a bound $M < \infty$ for $\left|\frac{\partial \tilde{v}}{\partial u}\right|$ in O. We cover C by open φ-squares and choose a finite subcovering $\{Q_j\}$. This covering is kept fixed. Let α_j be the horizontal cross cut of Q_j through its center. We set $M_1 = \sum_j a_j$, where a_j is the φ-length of α_j. We now choose, for given $\varepsilon_1 > 0$, a subinterval α of α_j of φ-length $a < \varepsilon_1$. We will first work with this α and sum up afterwards. We recall the following facts from Section 13.6: the set of vertical spirals through α is called a spiral set $\mathscr{S}(\alpha)$. It can of course be empty. The intersection $E = \alpha \cap \mathscr{S}(\alpha)$ is measurable. $\mathscr{S}(\alpha)$ can be partitioned into strips S_ν of vertical intervals (of equal φ-length) of the first kind and pairs of strips S_μ^+, S_μ^- of the second kind, of equal horizontal measure a_μ. As before, we can close any vertical interval β_ν in S_ν by a subinterval of α, and any pair β_μ^+ and β_μ^- of vertical intervals of S_μ^+ and S_μ^- respectively by two intervals on α. The φ-heights of these closed loops are b_ν in the first case, $b_\mu^+ + b_\mu^-$ in the second case. The heights with respect to $|d\tilde{v}|$ satisfy

$$b_\nu \le h_{\tilde{v}}(\gamma) \le \int_{\beta_\nu} \left|\frac{\partial \tilde{v}(u,v)}{\partial v}\right| dv + M \cdot \varepsilon_1. \tag{3}$$

There can be exceptional points in the strips S, but since they only accumulate to the ideal boundary of R, we can integrate (3) and get

$$a_\nu \cdot b_\nu \le \iint_{S_\nu} \left|\frac{\partial \tilde{v}(u,v)}{\partial v}\right| du\, dv + M \cdot \varepsilon_1 \cdot a_\nu, \tag{4}$$

where a_ν is the horizontal φ-measure of S_ν.

Let now S_μ^+ and S_μ^- be a pair of strips of the second kind of the decomposition of $\mathscr{S}(\alpha)$. We have, for the corresponding homotopy class of loops γ

$$b_\mu^+ + b_\mu^- \le h_{\tilde{v}}(\gamma) \le \int_{\beta_\mu^+} \left|\frac{\partial \tilde{v}}{\partial v}\right| dv + \int_{\beta_\mu^-} \left|\frac{\partial \tilde{v}}{\partial v}\right| dv + 2M\varepsilon_1. \tag{5}$$

This is true for every vertical interval β_μ^+ and β_μ^- (all the curves γ we can form in this way belong to the same homotopy class). Therefore we can take the infimum in both integrals, multiply with a_μ and then replace the products on the right hand side by double integrals:

$$\begin{aligned} a_\mu(b_\mu^+ + b_\mu^-) &\le a_\mu \cdot \inf_{\{\beta_\mu\}} \int_{\beta_\mu} \left|\frac{\partial \tilde{v}}{\partial v}\right| dv + a_\mu \inf_{\{\beta_\mu^-\}} \int_{\beta_\mu^-} \left|\frac{\partial \tilde{v}}{\partial v}\right| dv + a_\mu 2M\varepsilon_1 \\ &\le \iint_{S_\mu^+} \left|\frac{\partial \tilde{v}}{\partial v}\right| du\, dv + \iint_{S_\mu^-} \left|\frac{\partial \tilde{v}}{\partial v}\right| du\, dv + a_\mu 2M\varepsilon_1. \end{aligned} \tag{6}$$

We can now sum up (4) and (6) over the strips determined by α to get

$$\|\varphi\|_{\mathscr{S}(\alpha)} \le \iint_{\mathscr{S}(\alpha)} \left|\frac{\partial \tilde{v}}{\partial v}\right| du\, dv + a \cdot M\varepsilon_1. \tag{7}$$

To find the desired estimate for the norm, we subdivide all the α_j into finitely many pieces, each of length $<\varepsilon_1$. We number the pieces in any fixed way and start with the spiral set determined by the first one, then take what is left of the spiral set of the second one (see end of Section 13.6) and so on. Adding up the ring domains and these finitely many spiral sets (together they cover C up to a set of measure zero because of the condition on the vertical cross cuts) we get

$$\|\varphi\| - \varepsilon \leq \iint_R \left|\frac{\partial \tilde{v}}{\partial v}\right| du\, dv + M_1 M \varepsilon_1. \tag{8}$$

By refining the subdivision of the α_j we can let $\varepsilon_1 \to 0$, without changing M_1 and M. Therefore we can drop the last term in (8). The resulting inequality is true for every $\varepsilon > 0$. This leads to

$$\|\varphi\| \leq \iint_R \left|\frac{\partial \tilde{v}}{\partial v}\right| du\, dv. \tag{9}$$

The rest of the proof is the same as in 24.2, including the uniqueness part.

To show the last part of the theorem we simply have to add the vertical rectangles determined by the trajectory structure of φ in our estimate.

Corollary 24.5. *Let φ be a holomorphic quadratic differential of finite norm on a parabolic Riemann surface R. Let $\{\tilde{v}\}$ be as in Theorem* 24.5. *Then* (2) *holds, with equality only if $\{\tilde{v}\} = \{v\}$.*

Note that for every φ the vertical cross cuts cover a set of measure zero only.

24.6 The heights theorem (Marden and Strebel [1]). The norm inequality of Theorem 24.2 etc. applies to a pair of quadratic differentials $\tilde{\varphi}$ and φ, if we set $\tilde{v} = \operatorname{Im} \tilde{\Phi}$, $\tilde{\Phi} = \int \sqrt{\tilde{\varphi}}\, dz$. The theorem then reads: let φ be a holomorphic quadratic differential of finite norm on a parabolic Riemann surface R. If, for an arbitrary quadratic differential $\tilde{\varphi}$ on R $h_{\tilde{\varphi}}(\gamma) \geq h_{\varphi}(\gamma)$ for every simple loop γ on R, $\|\tilde{\varphi}\| \geq \|\varphi\|$, with equality only if $\tilde{\varphi} = \varphi$. For, in the case of equality, the function elements $\tilde{v}$ satisfy $\tilde{v} = \pm v + \text{const.}$, hence $\tilde{\Phi} = \pm \Phi + \text{const.}$ and consequently $\tilde{\varphi} = \varphi$. The most important consequence of this is the following uniqueness theorem.

Theorem 24.6 (The Heights Theorem). *Let φ and ψ be holomorphic quadratic differentials with finite norm on a compact Riemann surface with punctures or, more generally, on a parabolic surface. Let $h_\varphi(\gamma) = h_\psi(\gamma)$ for every simple closed loop γ on R. Then $\varphi = \psi$.*

The proof is immediate: we have $\|\psi\| \geq \|\varphi\|$ and $\|\varphi\| \geq \|\psi\|$, hence $\|\varphi\| = \|\psi\|$, and from the statement about equality we get $\varphi = \psi$.

The theorem can of course again be generalized to bordered surfaces with a parabolic double across the border. The heights must then also be the same for arcs joining boundary intervals of the surface. Vertical boundary intervals (for both differentials) can be neglected: for arcs joining vertical boundary curves heights need not be equal, because they do not go into the computation.

24.7 Continuity of the heights. We consider a sequence of holomorphic quadratic differentials φ_n on an arbitrary Riemann surface R which converges locally uniformly to a quadratic differential φ. Let γ be a simple closed loop on R, or even an arbitrary closed curve. We can choose, for given $\varepsilon>0$, a rectifiable curve γ_0 in the free homotopy class of γ, avoiding the zeroes of φ, and such that $\int_{\gamma_0}|dv|<h_\varphi(\gamma)+\varepsilon$. Of course $h_{\varphi_n}(\gamma)\leq\int_{\gamma_0}|dv_n|$. Since the sequence (φ_n) converges uniformly on γ_0 to φ, we have $\int_{\gamma_0}|dv_n|\to\int_{\gamma_0}|dv|$, and therefore $\overline{\lim_{n\to\infty}}\, h_{\varphi_n}(\gamma)\leq h_\varphi(\gamma)$. If $h_\varphi(\gamma)=0$, we end up with convergence $h_{\varphi_n}(\gamma)\to h_\varphi(\gamma)$. It is the contents of this section to show that this is true without any restriction.

Theorem 24.7. *Let (φ_n) be a sequence of holomorphic quadratic differentials on an arbitrary Riemann surface R which converges locally uniformly to a quadratic differential φ. Then, for every closed loop γ we have*

$$\lim_{n\to\infty} h_{\varphi_n}(\gamma)=h_\varphi(\gamma).$$

Note that the norms do not have to be finite and the surface R is arbitrary in this theorem.

Proof. The proof is in several steps.

(1) We start with a sequence of holomorphic quadratic differentials φ_n on an arbitrary Riemann surface R, tending locally uniformly to a differential φ, and a homotopically non trivial closed curve γ. Let $\hat{R}(\gamma)$ be the annular covering surface of R associated with γ (Section 2.2). $\hat{R}(\gamma)$ is represented by a circular annulus $0\leq r_0<|\zeta|<r_1\leq\infty$. The curve γ has a closed lift $\hat{\gamma}$ to $\hat{R}(\gamma)$ with winding number one around zero. The same is true for every closed curve γ_1 in the free homotopy class of γ. Conversely, the projection γ_1 of every curve $\hat{\gamma}_1$ with winding number one around zero is freely homotopic to γ. Moreover, if the curves are rectifiable and if $\hat{\varphi}$ denotes the lift of φ to $\hat{R}(\gamma)$, we have, with the usual notation, $\int_\gamma|dv|=\int_{\hat{\gamma}}|d\hat{v}|$. Therefore $h_\varphi(\gamma)=h_{\hat{\varphi}}(\hat{\gamma})$. We can thus, from the very beginning, start with an annular surface R and the class of closed curves with winding number one around zero. Moreover, we assume $\varphi\not\equiv 0$, because if $\varphi\equiv 0$ the theorem is already shown in the introduction.

(2) For a given positive number ε let γ be a curve with winding number one around zero and such that $\int_\gamma|dv|<h_\varphi(\gamma)+\varepsilon$. We choose γ to be a φ-polygon, composed of horizontal and vertical sides α_j, β_j and not passing through a zero of φ (Section 24.1). We can assume that γ is simple, because otherwise we could shorten it.

Let α_{jk} be a regular horizontal arc joining an interior point of β_j to an interior point of β_k but otherwise lying inside of γ (Fig. 69). It can then be extended to a maximal open horizontal strip S_{jk} inside of γ which has its two vertical sides γ_j and γ_k on β_j and β_k respectively. The φ-lengths of γ_j and γ_k are equal to the height of the strip S_{jk}.

The number of strips is bounded by the number of sides of γ. First, assume that $j=k$. The horizontal arc α_{jj} has both its endpoints on β_j and lies otherwise in the interior of γ. It therefore splits this interior into two Jordan domains,

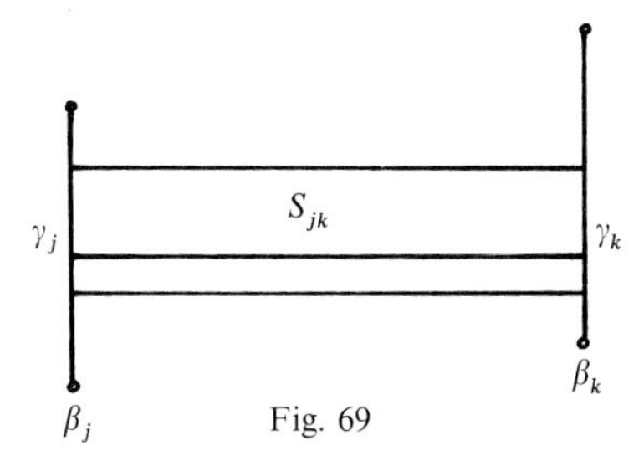

Fig. 69

one of which is bounded by α_{jj} and a subinterval of β_j. Because of the uniqueness theorem for shortest connections this domain must contain the inner boundary component of R. Let $\tilde{\alpha}_{jj}$ be another horizontal arc inside of γ with its two endpoints on β_j. Then the two arcs α_{jj} and $\tilde{\alpha}_{jj}$ bound, together with the two connecting subintervals on β_j, a simply connected domain. Because of the divergence principle, this must be a φ-rectangle, swept out by horizontal arcs. Therefore the associated horizontal strips S_{jj} and $\tilde{S}_{jj}$ coincide. We conclude that there can be at most one interior strip connecting β_j with itself.

Let now $j \neq k$. Then there can be two interior strips S_{jk}. But the two must be separated by the inner boundary component of R, and a similar argument shows that there cannot be more than two strips. There is no need to choose different notations for the (possibly) two of them, so we just speak of the strip (strips) S_{jk}.

Since the number of sides of γ is finite, and γ is kept fixed, the number of strips (interior and exterior) is bounded.

(3) Let us look at the interior strips S_{jk} with their heights d_{jk}. We need an upper estimate for the sum $\sum d_{jk}$. To this end, fix a strip S_{jk} and consider a horizontal arc α_{jk} in S_{jk}. It cuts off (i.e. separates from the inner boundary component of R) a certain subarc σ_{jk} of γ. It is easy to see that two such subarcs σ_{jk} and $\sigma_{h\ell}$ are either disjoint or one is contained in the other. There are thus finitely many maximal open subarcs σ_{jk} of γ, determined by the corresponding strips S_{jk}.

Let β_ℓ be a vertical side of σ_{jk}. A horizontal arc α leaving β_ℓ towards the interior of γ clearly must cut σ_{jk} again. It is therefore inside or on the boundary of a horizontal strip $S_{h\ell}$ with both its vertical sides on σ_{jk}. The total length of the vertical sides of all the maximal segments σ_{jk} is therefore equal to $2\sum d_{h\ell}$, where the sum goes over all the interior strips of γ.

On the other hand, the segment σ_{jk} can be cut off by a horizontal interval α_{jk} in S_{jk} (Fig. 70). Actually we can choose the endpoints of α_{jk} as close to the endpoints of the open arc σ_{jk} as we want. The height of the new curve is of course still greater or equal to $h(\gamma)$. We therefore get

$$\sum b_j - 2\sum d_{h\ell} \geq h(\gamma) > \sum b_j - \varepsilon.$$

We conclude that

$$2\sum d_{h\ell} < \varepsilon,$$

i.e. twice the sum of the heights of the inner strips is at most ε. The same is true for the sum of the heights of the horizontal strips in the exterior of γ.

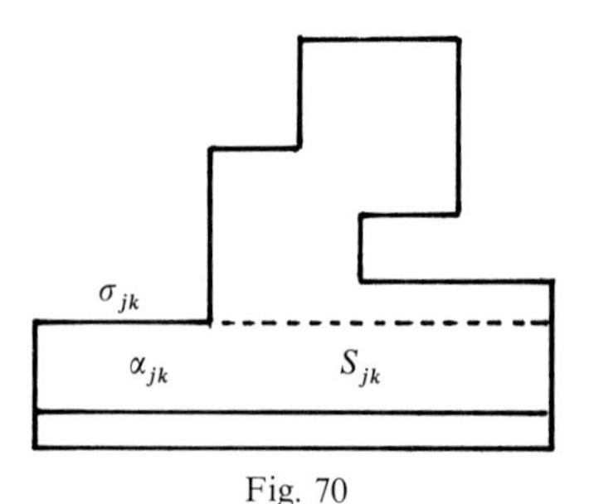

Fig. 70

(4) The next step is a lower estimate of the height $h(\gamma)$ by means of the sum of the heights of the strips. Fix the attention to the closures τ_i of the vertical sides of the horizontal strips. This is a finite collection of closed subintervals of the sides β_j of γ. A regular horizontal trajectory through a point $P \in \gamma \setminus \bigcup \tau_i$ cannot cut γ again and is therefore a "radial" cross cut of R. i.e. a trajectory tending to the inner boundary component of R in one, to the outer boundary component in the other direction. Let now γ_1 be a rectifiable closed curve in R with winding number one around zero. Just as in the proof of inequality (4) of Theorem 24.1 we can show that

$$\int_{\gamma_1} |dv| \geq \sum b_j - \sum |\tau_i|,$$

and since $\sum |\tau_i| = 2\sum d_{jk} + 2\sum d'_{jk}$ with d_{jk} and d'_{jk} the heights of the inner and outer strips S_{jk}, S'_{jk} respectively. We conclude that

$$h(\gamma) \geq \sum b_j - 2\sum d_{jk} - 2\sum d'_{jk},$$

which is the desired estimate.

(5) Consider the sequence (φ_n). Because of its locally uniform convergence to φ and the compactness of the curve γ it is clear that there exists a sequence of curves γ_n with the following properties: γ_n is a φ_n-polygon consisting of horizontal and vertical sides. Not only the sequence (γ_n) tends to γ, but the sides α_{jn}, β_{jn} of γ_n correspond to the sides α_j, β_j of γ and their lengths a_{jn}, b_{jn} tend to a_j, b_j respectively. Moreover, if S_{jk} is a horizontal strip of φ joining β_j and β_k, there is (for all sufficiently large n) a strip S_{jkn} of φ_n joining β_{jn} to β_{kn}. Also, the heights of the strips satisfy $\varliminf_{n\to\infty} d_{jkn} \geq d_{jk}$. But it is actually the converse inequality that we need.

Let S_{jkn} be a sequence of horizontal strips with respect to φ_n, joining β_{jn} and β_{kn}. Moreover, let the heights $d_{jkn} \to d > 0$, and let the vertical sides $\tau_{jn} \to \tau_j \subset \beta_j$, $\tau_{kn} \to \tau_k \subset \beta_k$. Let α_{jkn} be a horizontal segment in S_{jkn} which subdivides the strip S_{jkn} in a certain fixed ratio q, independent of n. We now have two strips, and one of them, S'_{jkn} say, is separated by the other, S''_{jkn}, from the inner boundary component of R. Since the moduli of the strips S''_{jkn} are bounded below and the vertical sides τ''_{jn} and τ''_{kn} tend to two intervals τ''_j and τ''_k respectively, the horizontal intervals α_{jkn} must stay away from the inner boundary component of R, i.e. in a compact subannulus of R. Because of the locally uniform convergence, the strips S'_{jkn} tend to strips S'_{jk} of φ, and this is true for every $q > 0$. We conclude that there is a strip S_{jk} with height $d_{jk} \geq d$.

Since we can start with a subsequence of the sequence (φ_n) from the beginning, we conclude in the usual way that the strips S_{jkn} converge to the strips S_{jk}, and in particular $d_{jkn} \to d_{jk}$.

(6) It is now easy to finish the proof. We have

$$h_{\varphi_n}(\gamma) \geq \sum b_{jn} - 2\sum d_{jkn} - 2\sum d'_{jkn},$$

hence

$$\varliminf_{n\to\infty} h_{\varphi_n}(\gamma) \geq \sum b_j - 2\sum d_{jk} - 2\sum d'_{jk} > \sum b_j - 2\varepsilon > h_\varphi(\gamma) - 2\varepsilon.$$

As $\varepsilon > 0$ is arbitrary, we have

$$\varliminf_{n\to\infty} h_{\varphi_n}(\gamma) \geq h_\varphi(\gamma),$$

which finishes the proof of the theorem.

§ 25. Approximation by Quadratic Differentials with Closed Trajectories

Introduction. Let us consider a recurrent trajectory ray α^+ which starts from a point P_0 on an arbitrarily short open vertical interval β. We orient β and assume that α^+ leaves it at the point P_0 to the right. If the first return point P_1 of α^+ on β is from the left, the interval $[P_0, P_1]$ on α^+ can be closed by the subinterval $[P_1, P_0]$ of β to form a simple loop (Fig. 71). Let α^+ pass β at P_1 from the right. Pick an open subinterval of β bounded by P_1 and a point P'_1 on the other side of P_0. If the first intersection point P_2 of α^+ with (P_1, P'_1) is from the left, we have the earlier situation. If it is from the right, the subinterval $[P_1, P_2]$ of α^+ together with its link on β forms a simple loop. Since β was arbitrary, the vertical link on β is arbitrarily short, i.e. the horizontal subinterval of α^+ is nearly closed.

It is this fact, together with the construction of quadratic differentials with closed trajectories as solutions of an extremal metric problem which makes it reasonable to think that the quadratic differentials with closed trajectories are

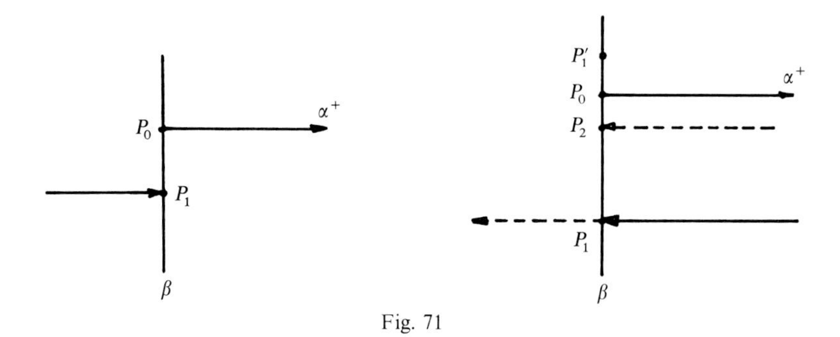

Fig. 71

dense in the whole space. A proof was given by the author in [8] for these cases, in which there exists a recurrent ray α^+ in each spiral set which has equally distributed intersections with the orthogonal interval β. "Equally distributed" means that the number of intersections of the interval $[P_0, P]$ of α^+ with any subinterval I of β divided by the total number of intersections with β tends to the quotient of the φ-lengths of I and β if P runs through α^+.

However, it was shown by H.B. Keynes and D. Newton in [1] that the intersections need not be equally distributed. A complete proof of the density theorem for holomorphic quadratic differentials on compact Riemann surfaces was then given by A. Douady and J. Hubbard in [1], using variational methods.

It is possible to construct an equally distributed admissible curve system by changing the surface slightly. This allows to apply the original idea of proof by the length area method (Strebel [12]). Using a construction devised by W. Thurston the different loops can be tied together to form a single Jordan curve on R. With this additional step one gets the stronger result, first shown by Masur in [1], that the differentials with a single cylinder (simple differentials) are dense in the whole space.

This line of proof does not make use of the heights theorem (Section 24.6). On the other hand, the simple differentials, together with intersection numbers, are conceptually very close to the idea of height. We therefore give a second proof of the approximation by simple differentials in the later part of this § using the heights theorem.

25.1 Quadratic differentials with a spiral which is dense on R. The case we treat here is that of a compact surface with finitely many punctures. In order to simplify the construction we first approximate φ by a quadratic differential with a trajectory structure consisting of a single spiral domain.

Theorem 25.1. *On a compact Riemann surface R with finitely many punctures P_j every holomorphic quadratic differential φ of finite norm can be approximated arbitrarily and in norm by quadratic differentials which have a spiral trajectory which is dense on R.*

Proof. The quadratic differential φ has at most poles of the first order at the points P_j. The boundary of a spiral domain consists of critical trajectories of finite length and their limiting critical endpoints (Section 11.2, Corollary 2). The same is of course true for the cylinders swept out by closed trajectories. Therefore, if there are no critical trajectories of finite length, there cannot be any ring domain and there is at most one spiral domain. As φ has finite norm, this must necessarily cover the whole surface.

We look at the differentials $\varphi_\vartheta = e^{-i\vartheta}\varphi$. For $\vartheta \to 0$, they approximate φ arbitrarily, since $\|\varphi_\vartheta - \varphi\| = |1 - e^{-i\vartheta}| \cdot \|\varphi\| \to 0$. The horizontal trajectories of φ_ϑ are the lines $e^{-i\vartheta}\varphi(z)\,dz^2 > 0$, i.e. $\arg\varphi(z)\,dz^2 = \vartheta$, whereas for the horizontal trajectories of φ we have $\arg\varphi(z)\,dz^2 = 0$.

We now have to choose ϑ in such a way that there is no line $\arg\varphi(z)\,dz^2 = \vartheta$ going from one critical point to another. Such a line is a geodesic and it is therefore the unique shortest connection of the two critical points in its homotopy class. There are only finitely many critical points, and for any two of them there are only denumerably many homotopy classes of arcs joining them (as one can see by triangulating the surface e.g.). Therefore the set of arguments ϑ for which there is an arc $\arg\varphi(z)dz^2=\vartheta$ joining two zeroes is denumerable, and in every neighborhood of $\vartheta=0$ there is a value $\vartheta\neq 0$ for which there is no such arc. The corresponding differential φ_ϑ has the desired property.

25.2 Approximation on compact surfaces with punctures

Theorem 25.2. *On a compact Riemann surface with finitely many punctures the quadratic differentials with closed trajectories are dense in the whole space of holomorphic quadratic differentials with finite norm.*

Proof. We give here a simplified version of the proof presented in [12]. It consists of three steps.

a) In the first step we construct a sequence of admissible systems of Jordan curves on $R=R\setminus\{P_j\}$ which are equally distributed in a sense to be precised. According to the preceding section we can start with a quadratic differential $\varphi \neq 0$ which has a spiral which is dense on R. We choose a closed vertical interval β and denote its φ-length by b. The rectangular strips of the first and second kind, with their vertical sides on β, are denoted generically by S_ν, their lengths and heights by a_ν and b_ν respectively. Since the vertical sides fill up both edges of β, we have $\sum b_\nu=b$.

We now construct a sequence of new surfaces R_n, which we call models, in the following way. We change the heights b_ν of the rectangles S_ν by less than $\frac{1}{n}$ such that they become all rational multiples of b, while keeping their lengths a_ν fixed. The new rectangles $S_{\nu n}$ are then identified along their horizontal sides in exactly the same way as the S_ν are sewn together on R. The $b_{\nu n}$ can clearly be chosen such that not only $\sum b_{\nu n}=b$ again, but also the vertices of the S_ν on β are shifted by less than $\frac{1}{n}$.

There is a piecewise affine mapping f_n of the model R_n onto R, in terms of the variable w, except for the points of β: f_n simply maps $S_{\nu n}$ onto S_ν by a vertical stretching.

The mapping can be extended to the two edges β^+ and β^- of β, which has to be looked at as slit. The extensions are piecewise linear, but in general they do not agree.

Let δ_n be the greatest common measure of all the heights $b_{\nu n}$. Each rectangle $S_{\nu n}$ is subdivided horizontally into $k_{\nu n}=\frac{b_{\nu n}}{\delta_n}$ congruent rectangles. The subrectangles clearly match together on β, and they can therefore be sewn along their vertical sides to form a certain number of long thin bands. Let the middle lines of these bands, separating their two boundary components, be denoted by $\tilde{\gamma}_{in}$. They form a finite system of disjoint Jordan curves on the model R_n. Going back to the original surface R by the mapping f_n, we find the curve system $\{\gamma_{in}\}$. Each γ_{in} is a loop consisting of horizontal intervals and vertical links of length less than $\frac{1}{n}$ on β. By taking the two edges of β slightly apart it is easy to see that the γ_{in} are freely homotopic to a set of disjoint Jordan curves on R. We do not claim that they all belong to different homotopy classes.

We introduce the following notations: $k_{i\nu n}$ is the number of times γ_{in} passes through S_ν. We then find

$$k_{\nu n} = \frac{b_{\nu n}}{\delta_n} = \sum_i k_{i\nu n}, \tag{1}$$

and for the total number of horizontal intervals

$$k_n = \frac{b}{\delta_n} = \sum_{i,\nu} k_{i\nu n}. \tag{2}$$

The horizontal length of γ_{in} is

$$A_{in} = \sum_\nu k_{i\nu n} \cdot a_\nu. \tag{3}$$

Finally, from $b_{\nu n} \to b_\nu$ and the relations (1) and (2) we conclude

$$\lim_{n\to\infty} \frac{1}{k_n} \sum_i k_{i\nu n} = \frac{b_\nu}{b}. \tag{4}$$

b) In the second step we construct an approximating sequence of quadratic differentials using the sequence of admissible curve systems $\{\gamma_{in}\}$, $n=1,2,3,\ldots$. Let φ_n be the solution of the extremal metric problem defined by the free homotopy classes of the curve system $\{\gamma_{in}\}$ with the lengths $\geq A_{in}$ (Section 21.13). The quadratic differential φ_n has closed trajectories homotopic to the curves γ_{in}. The lengths of the trajectories are $a_{in} = A_{in}$, because we only have to consider the non degenerate cylinders. As the quadratic differential φ itself satisfies the condition (Theorem 18.2), we have $\|\varphi_n\| \leq \|\varphi\|$. Therefore there exists a subsequence, which we can denote by (φ_n) again, which converges locally uniformly on $\dot{R}$ to a quadratic differential φ_0. It is easy to see, using square roots as parameters near the punctures, that the convergence is in norm. Clearly $\|\varphi_0\| \leq \|\varphi\|$.

c) The next and final step consists in showing that $\varphi_0 = \varphi$. The closure of the strip S_ν is represented as a horizontal rectangle $u_1 \leq u \leq u_2 = u_1 + a_\nu$, $v_1 \leq v \leq v_2 = v_1 + b_\nu$, in the $w = u + iv$-plane, $w = \Phi(z)$. Let $w_0 = u_0 + iv_0 = \Phi_0(z) = \int \sqrt{\varphi_0(z)}\, dz$. The integral

$$\int |du_0| = \int_{u_1}^{u_2} \left| \frac{\partial u_0(u,v)}{\partial u} \right| du \tag{5}$$

is a continuous function of v in the closed interval $v_1 \leq v \leq v_2$. This is evident at the interior points of the interval $[v_1, v_2]$, because there, the integrand is continuous, except at the possible zeroes of φ_0. Moreover, it is dominated by

$$\left|\frac{\partial u_0}{\partial u}\right| \le \left|\frac{dw_0}{dw}\right| = |\varphi_0(w)|^{\frac{1}{2}}. \tag{6}$$

Therefore, the integral (5), over an arbitrary small horizontal interval I, is dominated by

$$\int_I |\varphi_0(w)|^{\frac{1}{2}} du, \tag{7}$$

which is the φ-length of I.

The lower and upper sides of S_ν can correspond to critical horizontal arcs on R. At a critical point P the function $w = \Phi(z)$ cannot be used as a surface parameter (Fig. 72). Let ζ be the natural parameter with respect to φ near P, ζ

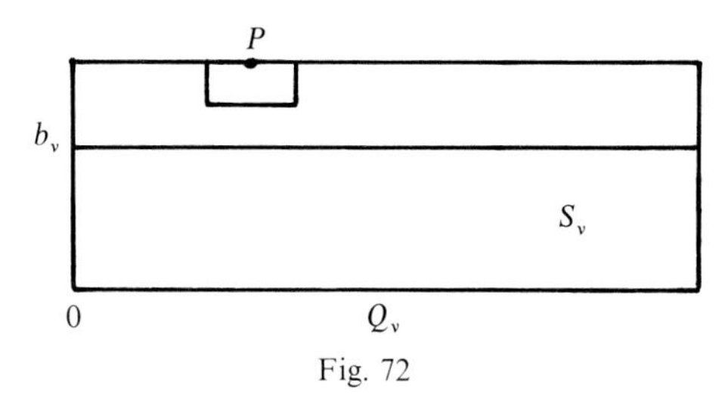

Fig. 72

$= 0 \leftrightarrow P$. We have $w = \Phi(\zeta) = \zeta^{\frac{n+2}{2}}$, $n \ge -1$ (see Theorem 6.2). The differential φ_0 has a representation, in terms of ζ, of the form

$$\varphi_0(\zeta) \sim \text{const.}\, \zeta^m, \qquad m \ge -1. \tag{8}$$

We therefore find

$$w_0 = \Phi_0(\zeta) \sim \text{const.}\, \zeta^{\frac{m+2}{2}} = \text{const.}\, w^{\frac{m+2}{n+2}}, \tag{9}$$

and

$$|\varphi_0(w)|^{\frac{1}{2}} \sim \text{const.}\, |w|^{\frac{m-n}{n+2}}. \tag{10}$$

The exponent is $\ge -\dfrac{n+1}{n+2} > -1$. We can therefore exclude the point P by a rectangular half neighborhood such that the integral (7) over any horizontal interval I in this neighborhood, is arbitrarily small. From this, the continuity of the integral (5) easily follows.

We now choose an arbitrary positive ε. We can then find a subdivision of the strips S_ν into finitely many horizontal substrips, which we may call S_ν again, and positive numbers c_ν, such that, for any horizontal in S_ν,

$$c_\nu - \varepsilon < \int \left|\frac{\partial u_0(u,v)}{\partial u}\right| du < c_\nu + \varepsilon. \tag{11}$$

The subdivision is merely for integration purposes and does not influence the sequence of curve systems $\{\gamma_{in}\}$, which is kept fixed. Of course the quantities with the index ν in the relations (1)–(4) are changed, i.e. b_ν, $b_{\nu n}$, $k_{\nu n}$, $k_{i\nu n}$, and a_ν as far as the numbering goes, but otherwise (1)–(4) stay intact. In particular, the horizontal lengths A_{in} of the curves γ_{in} are of course the same, therefore φ_0 is untouched. Since $\varphi_n \to \varphi_0$ locally uniformly on R, inequality (11) also holds with u_n instead of u_0, i.e.

$$c_\nu - \varepsilon < \int \left| \frac{\partial u_n(u, v)}{\partial u} \right| du < c_\nu + \varepsilon, \tag{12}$$

for all sufficiently large n.

For every ring domain of φ_n the length a_{in} of the closed trajectories is equal to the given quantity A_{in}. This is therefore also the minimum of the total variation of u_n in the corresponding free homotopy class. In particular

$$a_{in} = \sum_\nu k_{i\nu n} a_\nu \leq \int_{\gamma_{in}} |du_n|. \tag{13}$$

The curve γ_{in} is composed of its horizontal intervals, in the rectangles S_ν, of length a_ν, and its vertical links on β, of length $< \frac{1}{n}$. Let M be a bound for $\left|\frac{\partial u_n}{\partial v}\right|$ on β, for all n. We then have, for all sufficiently large n,

$$\int_{\gamma_{in}} |du_n| < \sum_\nu k_{i\nu n}(c_\nu + \varepsilon) + \varepsilon \cdot M \sum_\nu k_{i\nu n} a_\nu. \tag{14}$$

Combining (13) and (14) we get

$$(1 - M\varepsilon) \sum_\nu k_{i\nu n} a_\nu \leq \sum_\nu k_{i\nu n}(c_\nu + \varepsilon). \tag{15}$$

Summing (15) over all i, dividing by k_n and using relation (4) leads to

$$\|\varphi\| = \sum a_\nu b_\nu \leq \sum_\nu b_\nu (c_\nu + \varepsilon)(1 - M\varepsilon)^{-1}. \tag{16}$$

Integrating (11) over v in S_ν and summing over the ν yields

$$\iint_R \left| \frac{\partial u_0(u, v)}{\partial u} \right| du\, dv > \sum b_\nu (c_\nu - \varepsilon) = \sum b_\nu c_\nu - b\varepsilon. \tag{17}$$

Combining (14) with (17) and letting $\varepsilon \to 0$ finally gives

$$\|\varphi\| \leq \iint_R \left| \frac{\partial u_0(u, v)}{\partial u} \right| du\, dv. \tag{18}$$

We now apply Schwarz's inequality and complete the right hand integral to the Dirichlet integral. We find

$$\begin{aligned} \|\varphi\|^2 &\leq \|\varphi\| \cdot \iint_R \left(\frac{\partial u_0}{\partial u} \right)^2 du\, dv \\ &\leq \|\varphi\| \cdot \iint_R \left\{ \left(\frac{\partial u_0}{\partial u} \right)^2 + \left(\frac{\partial u_0}{\partial v} \right)^2 \right\} du\, dv = \|\varphi\| \cdot \|\varphi_0\|. \end{aligned} \tag{19}$$

As $\|\varphi_0\| \leq \|\varphi\|$, we must have equality. We conclude, as in Section 24.2 after inequality (14), but with u_0 and u instead of $\tilde{v}$ and v, that $u_0 = \pm u + \text{const.}$, hence $\Phi_0 = \pm \Phi + \text{const.}$, $\varphi_0 = \varphi$.

The following slight generalization is possible. Let R be a compact surface with border, let φ be meromorphic on R with finite norm, and positive along the boundary components, except for possible zeroes. Denote the possible first order poles of φ in the interior of R by P_j. Then φ can be approximated by quadratic differentials which are holomorphic on $\dot{R}=R\setminus\{P_j\}$, have finite norm and closed trajectories.

In the proof we cannot make use of a rotation of φ to $\varphi_\vartheta=e^{-i\vartheta}\varphi$, because of the boundary of R. The trajectory structure now consists, in general, of finitely many ring domains of closed trajectories and finitely many spiral domains, the number of critical trajectories and cross cuts being finite. We choose a vertical arc in every spiral domain. This leads to a system of simple loops, in every spiral domain, as above. The system is completed by a set of closed trajectories, one out of every ring domain. We can now construct the quadratic differentials φ_n, as before. To arrive at the estimate (18), we have to apply the integration procedure to every ring domain and every spiral domain and add up. The proof is then finished as before.

25.3 Approximation by simple differentials. We return to the case of a compact surface R without border. Let φ be meromorphic on R, with finite norm, and with a spiral which is dense on R. According to 25.1 we can choose the vertical interval β such that its two endpoints are connected by a horizontal interval α which hits β from two different sides. We now pass to the model R_n as described in Section 25.3(a). We have the closed bands R_{in} and their middle lines $\tilde{\gamma}_{in}$, as constructed there. The $\tilde{\gamma}_{in}$ are simple loops. Following an idea of Thurston we are going to replace them by parallel strands of simple loops which can then in turn be connected, using a twisting, to form a single simple loop.

Let $i(\tilde{\gamma}_{in},\beta)$ be the intersection number of $\tilde{\gamma}_{in}$ with β. It is simply the number of common points of $\tilde{\gamma}_{in}$ and β, which is the same as the number of times R_{in} passes through β. R_{in} can be mapped conformally onto a circular annulus in the plane, its $i(\tilde{\gamma}_{in},\beta)$ intersections with β being represented by radii. The simplest way to incorporate $\tilde{\gamma}_{in}$, together with $\sigma=\alpha+\beta$, into a loop, would be: follow σ until it meets $\tilde{\gamma}_{in}$, go around $\tilde{\gamma}_{in}$ and then continue along σ. But this only works in case $i(\tilde{\gamma}_{in},\beta)=1$; otherwise the resulting closed curve is not simple, because it intersects the other cross cuts of β in R_{in}.

In order to avoid that, we take $i(\tilde{\gamma}_{in},\beta)$ parallel strands of $\tilde{\gamma}_{in}$ in R_{in}, represented by concentric circles (Fig. 73). The rule is now the following: Enter R_{in} along σ; turn left on the intersection with the first circular arc; turn right on the next intersection with a radius. We end up on the same radius, leaving R_{in} through the other boundary component. The same process is repeated, on our way through σ, whenever we enter any R_{in}. It does not matter through which boundary component of R_{in} this happens. In the end we have used up all the strands and all the radii in all the R_{in}, all of them exactly once, and we clearly have a curve which is homotopic (by slight changes at the corners) to a simple closed loop. The horizontal length of the curve is equal to the sum of the lengths of the strands plus the length of the horizontal arc α connecting the two endpoints of β (the horizontal subarc of σ).

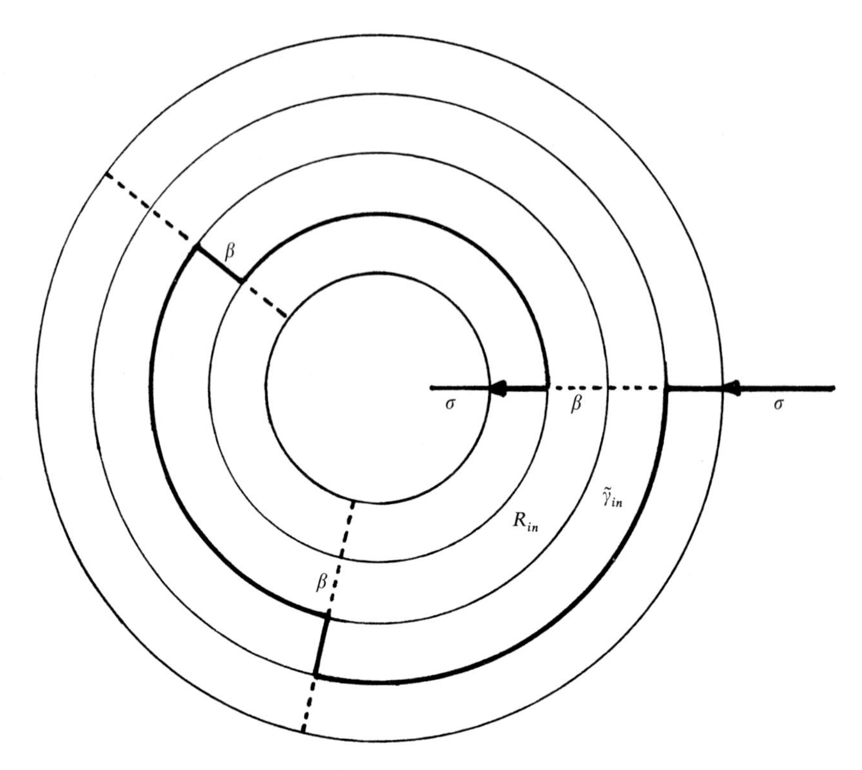

Fig. 73: R_{in}, $i(\tilde{\gamma}_{in}, \beta)=3$

Let now M_n be the least common multiple of the intersection numbers $i(\tilde{\gamma}_{in}, \beta)$ for all indices i. Putting M_n strands into each R_{in} instead of just $i(\tilde{\gamma}_{in}, \beta)$ we perform the same construction. It amounts to a twist of one boundary component of R_{in} with respect to the other by $2\pi M_n/i(\tilde{\gamma}_{in}, \beta)$ instead of 2π as in the first case (Dehn twist). The effect is that the resulting simple loop $\tilde{\gamma}_n$ now has the same number of strands in every cylinder R_{in}, in other words, it is equally distributed in the R_{in}.

We now go back from the model R_n to the surface R, as in Section 25.3 (a). The piecewise affine mapping f_n takes $\tilde{\gamma}_n$ into a loop γ_n on R which consists of horizontal segments of length a_ν in the strips S_ν and vertical links on β in between, plus the fixed horizontal interval α connecting both ends of β.

The curve γ_n has, with the notation of Section 25.3 (a),

$$k_{\nu n} = \frac{b_{\nu n}}{\delta n} M_n \tag{1}$$

horizontal intervals in $S_{\nu n}$, hence in S_ν. Every vertical link on β is preceded and followed by a horizontal interval of length a_ν, for some ν. With increasing n, its length tends to zero, since the shift on β tends to zero as well as δ_n. The simple differential φ_n is constructed with closed trajectories α_n of homotopy type

γ_n and length

(2) $$a_n = \sum_{\nu} k_{\nu n} a_{\nu}.$$

The differential φ has lengths $\geq a_n + a$, with $a = |\alpha|_{\varphi}$ the length of the horizontal segment of σ. It therefore satisfies the length inequality and we have $\|\varphi_0\| \leq \|\varphi\|$. By an estimate from below we can show, in the very same way as in Section 25.3(c) – with the exception that there is no summation over i, because there is only one loop γ_n – that $\varphi_0 = \varphi$.

25.4 Intersection numbers. On an arbitrary Riemann surface R the universal covering surface of which is the disk we consider simple closed loops γ and σ which do not bound a disk or a once punctured disk on R.

Definition 25.4. The intersection number $i(\sigma, \gamma)$ of the two curves σ and γ is the minimum of points which any two simple loops σ_0 and γ_0 of the free homotopy classes of σ and γ respectively have in common.

Choose a point $P \in \gamma$ and a point $\hat{P} \in \hat{R}$ above P. Let $\hat{\gamma}$ be the lift of γ with initial point $\hat{P}$. It ends at a point $\hat{P}_1$ above P, and the cover transformation T of $\hat{R}$ which takes $\hat{P}$ into $\hat{P}_1$ is hyperbolic. The Poincaré geodesic connecting its two fixed points on $\partial \hat{R}$ is invariant under T. It projects onto the unique shortest loop γ_0 in the Poincaré metric on R which is freely homotopic to γ.

Let σ_0 be the closed Poincaré geodesic on R in the free homotopy class of σ. We now have two simple loops σ_0 and γ_0 in the free homotopy classes of σ and γ respectively, and therefore by definition

(1) $$i(\sigma, \gamma) \leq \operatorname{card} \sigma_0 \cap \gamma_0.$$

It follows easily that we have in fact equality. To this end, let σ_1 and γ_1 be simple closed loops freely homotopic to σ_0 and γ_0 respectively. Look at the loops as periodic curves on R. Let $\hat{\gamma}_0$ be a lift of γ_0. It is an orthogonal circle in $\hat{R}$. The curve γ_1 has a lift $\hat{\gamma}_1$ with the same endpoints on $\partial \hat{R}$. The lifts of σ_0 from its intersection points with γ_0 are different orthogonal circles of $\hat{R}$. The curve σ_1 has as many lifts, by homotopy, with the same respective endpoints. Therefore their number of intersections with $\hat{\gamma}_1$ is at least card $\sigma_0 \cap \gamma_0$. They project into necessarily different points of $\sigma_1 \cap \gamma_1$. Therefore $\operatorname{card} \sigma_1 \cap \gamma_1 \geq \operatorname{card} \sigma_0 \cap \gamma_0$, and since σ_1 and γ_1 are arbitrary

(2) $$i(\sigma, \gamma) \geq \operatorname{card} \sigma_0 \cap \gamma_0,$$

which proves the equality.

The intersection number of a closed loop σ with a closed trajectory α of a simple differential φ is in a natural way connected with the height of σ with respect to φ.

Theorem 25.4. *Let γ be a Jordan curve on R which does not bound a disk or a once punctured disk. Let $\varphi[\gamma]$ be a simple differential on R the closed trajectories of which are freely homotopic to γ. Then*

(3) $$h_{\varphi[\gamma]}(\sigma) = b\, i(\sigma, \gamma),$$

where b is the height of the cylinder of $\varphi[\gamma]$.

Proof. Take an arbitrary simple loop σ in a fixed free homotopy class. Replace it by a simple φ-polygon $\tilde{\sigma}$ in the same class and such that $\int_{\tilde{\sigma}} |dv| \le \int_{\sigma} |dv|$, with $w = u + iv = \Phi(z) = \int \sqrt{\varphi[\gamma]}\, dz$. Choose a closed trajectory α in the characteristic ring domain R_0 of φ. Any subarc of $\tilde{\sigma}$ which is a cross cut of R_0 either has both its endpoints on the same boundary component of R_0 or else it traverses R_0. Subarcs of the first kind can clearly be pushed to the boundary so that they do not meet α; subarcs of the second kind can be changed in such a way that they cut α only once. The integral $\int |dv|$ over such an arc is $\ge b$. We therefore have

$$\int_{\sigma} |dv| \ge \int_{\tilde{\sigma}} |dv| \ge i(\sigma, \gamma) \cdot b. \tag{4}$$

Since this is true for every σ of the given free homotopy class, we have

$$h_{\varphi[\gamma]}(\sigma) \ge i(\sigma, \gamma) \cdot b. \tag{5}$$

In order to prove the opposite inequality we choose the simple φ-polygon $\tilde{\sigma}$ in the free homotopy class of σ such that $i(\sigma, \gamma) = \operatorname{card} \tilde{\sigma} \cap \alpha$. Every subarc of $\tilde{\sigma}$ which is a cross cut of R_0 connecting the two different boundary components of R_0 intersects α once. It can clearly be replaced by a monotonous arc of height b. The finitely many remaining arcs have both their endpoints on the same boundary component of R_0. They can be replaced by arcs going along the boundary of R_0 of arbitrarily small height. Therefore, for the modified $\tilde{\sigma}$,

$$\begin{aligned} h_{\varphi[\gamma]}(\sigma) \le \int_{\tilde{\sigma}} |dv| &\le b \cdot \operatorname{card} \tilde{\sigma} \cap \alpha + \varepsilon \\ &= b \cdot i(\sigma, \alpha) + \varepsilon. \end{aligned} \tag{6}$$

We conclude that

$$h_{\varphi[\gamma]} \le b \cdot i(\sigma, \alpha) \tag{7}$$

which, together with (5) proves the equality.

25.5 The mapping by heights. The approximation by simple differentials in connection with the heights theorem enables us to establish an interesting correspondence between the quadratic differentials of two surfaces.

Theorem 25.5. *Let R and R' be two compact Riemann surfaces with finitely many punctures $\{P_j\}$ and $\{P_j'\}$ respectively. Then, any homeomorphism $f\colon R \to R'$ with $f(P_j) = P_j'$ for all j induces a homeomorphism of the space of holomorphic quadratic differentials of finite norm on $\dot{R} = R \setminus \{P_j\}$ onto the same space on $\dot{R}' = R \setminus \{P_j'\}$ by the requirement that the heights are the same for all pairs of corresponding simple loops.*

Let us denote the postulated mapping of the quadratic differentials by F. We thus have

$$F\colon \varphi \to \psi \quad \text{where} \quad h_{\varphi}(\sigma) = h_{\psi}(f(\sigma)) \tag{1}$$

for all simple loops σ on R.

Proof. The uniqueness of ψ for given φ follows from the heights theorem. For, let ψ and $\tilde{\psi}$ be two differentials corresponding to φ. Then, $h_\psi(\sigma')=h_{\tilde{\psi}}(\sigma')=h_\varphi(f^{-1}(\sigma))$ for every simple loop σ'. Therefore $\psi=\tilde{\psi}$. In particular $\varphi=0\mapsto\psi=0$.

We now show that the mapping exists. Let $\varphi\neq 0$ be an arbitrary holomorphic quadratic differential with finite norm on $\dot{R}$. Let (φ_n) be a sequence of simple differentials converging locally uniformly and hence in norm to φ. According to Theorem 25.5 we have

$$h_{\varphi_n}(\sigma)=b_n i(\sigma,\alpha_n) \tag{2}$$

for every simple loop σ, where α_n is a closed trajectory of φ_n and b_n is the height of its cylinder. Because of Theorem 24.7 (the proof of which could be very much simplified in the present case of a compact surface with punctures) $h_{\varphi_n}(\sigma)\to h_\varphi(\sigma)$. Denote by ψ_n the simple differential on R' with closed trajectories $\alpha'_n\sim f(\alpha_n)$ and height $b'_n=b_n$ of its cylinder R'_n. Theorem 21.1 ensures the existence of ψ_n. Since intersection numbers are preserved by f, we have $i(\sigma,\alpha_n)=i(f(\sigma),f(\alpha_n))$ and hence

$$h_{\psi_n}(f(\sigma))=b'_n i(f(\sigma),\alpha'_n)=b_n i(\sigma,\alpha_n)=h_{\varphi_n}(\sigma) \tag{3}$$

for every simple loop σ.

In order to show that the sequence (ψ_n) converges, we need the following

Lemma 25.5. *A set of holomorphic quadratic differentials of finite norm on a compact Riemann surface with punctures the heights of which are bounded is bounded in norm.*

Proof of the lemma. Let $h_\varphi(\sigma)\leq M(\sigma)$ for all simple loops σ on $\dot{R}$ and all quadratic differentials of the set. Note that the bound may depend on σ. Assume that the set is unbounded and let (φ_ν) be a sequence of differentials of the set with $\|\varphi_\nu\|\to\infty$. The differentials $\tilde{\varphi}_\nu=\dfrac{\varphi_\nu}{\|\varphi_\nu\|}$ have norm one. Therefore there exists a subsequence, which we denote by $(\tilde{\varphi}_\nu)$ again, which converges locally uniformly and, because $\dot{R}=R\setminus\{P_j\}$ and R is compact, even in norm to a quadratic differential $\tilde{\varphi}$. Of course, $\tilde{\varphi}$ is again holomorphic on $\dot{R}$ and has norm $\|\tilde{\varphi}\|=1$. Now $h_{\tilde{\varphi}_\nu}(\sigma)\to h_{\tilde{\varphi}}(\sigma)$ for every simple loop σ. Choose σ such that $h_{\tilde{\varphi}}(\sigma)>0$. Then

$$h_{\tilde{\varphi}_\nu}(\sigma)=\frac{1}{\|\varphi_\nu\|}h_{\varphi_\nu}(\sigma)\leq\frac{M(\sigma)}{\|\varphi_\nu\|} \tag{4}$$

and thus

$$\overline{\lim_{\nu\to\infty}}\,\|\varphi_\nu\|\leq\frac{M(\sigma)}{h_{\tilde{\varphi}}(\sigma)}<\infty \tag{5}$$

which contradicts the assumption.

Corollary of the lemma. Let (φ_n) be a sequence of holomorphic quadratic differentials of finite norm on $\dot{R}$ the heights $h_{\varphi_n}(\sigma)$ of which converge for every simple loop σ on $\dot{R}$. Then the sequence (φ_n) converges in norm to a differential φ with $h_\varphi(\sigma) = \lim_{n\to\infty} h_{\varphi_n}(\sigma)$ for all σ.

The proof of the corollary is immediate by the lemma and the heights theorem. For the sequence (φ_n) is bounded in norm. Therefore there exists a subsequence (φ_{n_i}) which converges in norm to a differential φ, $h_\varphi(\sigma) = \lim_{i\to\infty} h_{\varphi_{n_i}}(\sigma)$ for every simple loop σ on $\dot{R}$. φ is uniquely determined by its heights. We conclude, by a standard argument, that the original sequence converges to φ.

The existence of $F(\varphi) = \psi$ now follows from (3). It is independent of the approximating sequence (φ_n) of φ.

The mapping F is bijective. For, given ψ on $\dot{R}'$, there is, by the above construction, a φ on $\dot{R}$ with $h_\varphi(\sigma) = h_\psi(\sigma')$, $\sigma = f^{-1}(\sigma')$. Evidently $\psi = F(\varphi)$. And for $\varphi_1 \neq \varphi_2$ we have $F(\varphi_1) \neq F(\varphi_2)$.

Furthermore, F is continuous: Let $(\varphi_n) \to \varphi$, $\psi = F(\varphi)$. Since $h_{\varphi_n}(\sigma) \to h_\varphi(\sigma)$ for every simple loop σ, the heights $h_{\psi_n}(\sigma') = h_{\varphi_n}(\sigma)$ are bounded for every $\sigma' = f(\sigma)$. We conclude that the sequence $\|\psi_n\|$ is bounded, and therefore there exists a convergent subsequence $(\psi_{n_i}) \to \tilde{\psi}$,

$$h_{\tilde{\psi}}(\sigma') = \lim_{i\to\infty} h_{\psi_{n_i}}(\sigma') = h_\varphi(\sigma). \tag{6}$$

As before we conclude that $\tilde{\psi} = \psi$ and that the original sequence (ψ_n) converges to ψ.

The inverse mapping F^{-1} is nothing but the mapping by heights from $\dot{R}'$ to $\dot{R}$, induced by the homeomorphism f^{-1}. Therefore it is also continuous, and we have shown that F is a homeomorphism from $\{\varphi\}$ onto $\{\psi\}$.

25.6 A second proof of the approximation by simple differentials. The proof of the approximation of quadratic differentials of finite norm by simple differentials can be shortened considerably if one uses the heights theorem.

Let R be a compact Riemann surface and let $\{P_j\}$ be a finite set of points on R. The quadratic differential φ is supposed to be holomorphic on the punctured surface $\dot{R} = R \setminus \{P_j\}$ and have finite norm. The punctures are then regular points of φ or first order poles. By Theorem 25.1 we may assume, without restricting the generality, that φ has a spiral trajectory which is dense on R.

We fix a closed vertical interval β on $\dot{R}$ which does not meet a zero of φ. The horizontal φ-rectangles on $\dot{R}$, based on β, are denoted by S_ν. The variations $\tilde{R}_n$ of R (the models) are composed of the rectangles $\tilde{S}_{\nu n}$ (see Sections 25.2(a) and 25.3), with the proper identifications, except on β, where we have a piecewise linear shift.

The surface R can also be visualized as collection of the horizontal (Euclidean) rectangles S_ν over the Φ-plane, with the imposed identifications on their circumferences. The affine mappings of the $\tilde{S}_{\nu n}$ onto the S_ν compose to piecewise affine mappings f_n of the surfaces $\tilde{R}_n$ onto R, slit along β (the two edges are mapped differently). The surfaces $\tilde{R}_n$ are also supposed to have punctures: they are the pre-images by the f_n of the punctures of R. They lie on the upper and lower borders of the rectangles $\tilde{S}_{\nu n}$. Fix a horizontal rectangle S on R containing β as its vertical line of symmetry. It is clearly possible to change the mappings f_n slightly in S by a proper shift of the endpoints of the horizontals on β (as indicated in Fig. 74) to become a homeomorphism of $\tilde{R}_n$ onto R.

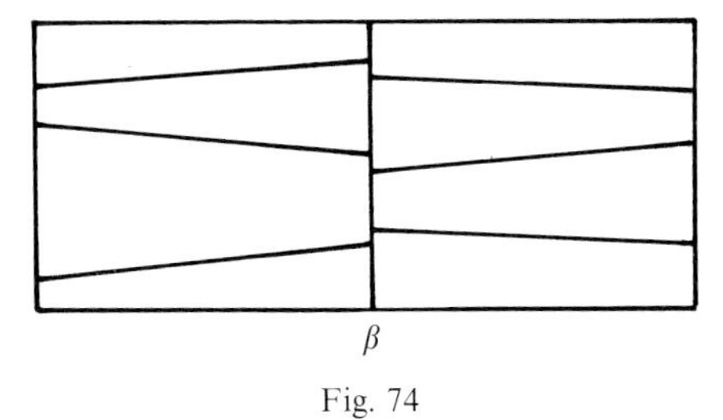

Fig. 74

This shows that the intersection number of any two simple loops σ and γ on R is the same as the intersection number of their images $\tilde{\sigma}_n$ and $\tilde{\gamma}_n$ on $\tilde{R}_n$. This remains true, if we denote by $\tilde{\sigma}_n$ and $\tilde{\gamma}_n$ the f_n^{-1} images of σ and γ respectively, with little vertical shifts on β, because the changes are clearly homotopies. We have

$$i(\sigma, \gamma) = i(\tilde{\sigma}_n, \tilde{\gamma}_n). \tag{1}$$

The differential φ on R is equal to one in terms of the parameter $w = \Phi(z)$ on the Euclidean rectangles S_ν. Similarly, on the model $\tilde{R}_n$, we have a quadratic differential $\tilde{\varphi}_n$, represented by the number one in the rectangles $\tilde{S}_{\nu n}$. This latter differential has closed trajectories, homotopic to the loops $\tilde{\gamma}_{in}$. Let σ be a simple loop on $\dot{R}$. We can start with the geodesic, on $\dot{R}$, in the free homotopy class of σ, and then replace it by a step curve, consisting of horizontal and vertical intervals. They can evidently be chosen such that σ has minimal height, being the sum of the lengths of its vertical intervals:

$$h_\varphi(\sigma) = \sum d_\mu. \tag{2}$$

The curve $\tilde{\sigma}_n$ on $\tilde{R}_n$ has height, with respect to $\tilde{\varphi}_n$,

$$h_{\tilde{\varphi}_n}(\tilde{\sigma}_n) \leq \sum \tilde{d}_{\mu n} + k\frac{2}{n}, \tag{3}$$

where k is the number of intersections of σ with β and the numbers $\tilde{d}_{\mu n} \leq d_\mu \left(1 + \frac{1}{n}\right)$ are the lengths of the vertical sides of $\tilde{\sigma}_n$. On the other hand, starting with a loop $\tilde{\sigma}_n$ on $\tilde{R}_n$, we can argue in the same way, but this time going from $\tilde{R}_n$ to R. Both inequalities together give the relation

$$\lim_{n \to \infty} h_{\tilde{\varphi}_n}(\tilde{\sigma}_n) = h_\varphi(\sigma). \tag{4}$$

The quadratic differential $\tilde{\varphi}_n$ has closed trajectories. The cylinders $\tilde{R}_{in}$ have all height δ_n. Therefore, by an argument similar to the one which led to Theorem 25.4, we find

$$h_{\tilde{\varphi}_n}(\tilde{\sigma}_n) = \delta_n \sum i(\tilde{\sigma}_n, \tilde{\gamma}_{in}), \tag{5}$$

where the sum goes over all simple loops $\tilde{\gamma}_{in}$ on $\tilde{R}_n$.

We now need an estimate of the number of intersections of Thurstons curve $\tilde{\gamma}_n$ with $\tilde{\sigma}_n$. It is given by the inequality

$$M_n \sum i(\tilde{\sigma}_n, \tilde{\gamma}_{in}) - i(\tilde{\sigma}_n, \tau) \leq i(\tilde{\sigma}_n, \tilde{\gamma}_n) \leq M_n \sum i(\tilde{\sigma}_n, \tilde{\gamma}_{in}) + i(\tilde{\sigma}_n, \tau),$$

where M_n is the number of strands parallel to $\tilde{\gamma}_{in}$ put into $\tilde{R}_{in}$ and τ is the transverse curve $\tau = \alpha + \beta$. (For the proof we refer to A. Fathi [3, p. 68] or to A. Marden and K. Strebel [1]).

The approximating simple differentials φ_n on R are constructed to have closed trajectories homotopic to γ_n and height of the cylinder $\frac{\delta_n}{M_n}$. Collecting the above inequalities we find

$$\begin{aligned} h_{\varphi_n}(\sigma) &= \frac{\delta_n}{M_n} i(\sigma, \gamma_n) = \frac{\delta_n}{M_n} i(\tilde{\sigma}_n, \tilde{\gamma}_n) \\ &= \frac{\delta_n}{M_n} \left(M_n \sum i(\tilde{\sigma}_n, \tilde{\gamma}_{in}) \pm i(\sigma, \tau)\right) \\ &= \delta_n \sum i(\tilde{\sigma}_n, \tilde{\gamma}_{in}) \pm \frac{\delta_n}{M_n} i(\sigma, \tau) \\ &= h_{\tilde{\varphi}_n}(\tilde{\sigma}_n) \pm \frac{\delta_n}{M_n} i(\sigma, \tau). \end{aligned} \tag{7}$$

Because of (4) and $\frac{\delta_n}{M_n} \to 0$ with $n \to \infty$ this tends to $h_\varphi(\sigma)$. Therefore by the corollary to Lemma 25.5 $\varphi_n \to \varphi$.

References

Ahlfors, L.V.: [1] Variational Methods in Function Theory. Lecture Notes, Harvard (1953).
- [2] On quasiconformal mappings. Journal d'Analyse Math. Vol. III (1953/54) 1-58.
- [3] Conformal invariants. McGraw-Hill Series in Higher Mathematics (1973) 1-157
- and Sario, L.: [1] Riemann surfaces. Princeton (1960) 1-382.

Carleman, T.: [1] Zur Theorie der Minimalflächen. Math. Zeitschr. 9 (1921) 154-160.

Douady, A. and Hubbard, J.: [1] On the density of Strebel differentials. Inventiones math. 30 (1975) 175-179.

Duren, P.L.: [1] Theory of H^p spaces. Academic Press (1970) 1-258.

Fathi, A., Laudenbach, F., Poénaru, V.: [1] Travaux de Thurston sur les surfaces. Séminaire Orsay, Astérisque 66-67 (1979) 1-284.

Gardiner, F.P.: [1] The existence of Jenkins-Strebel differentials from Teichmüller theory. Amer. J. of Math. 99 No. 4 (1977) 1097-1104.
- [2] On Jenkins-Strebel differentials for open Riemann surfaces. Proceedings of the Royal Soc. of Edinburgh 86 A (1980) 315-325.

Goodman, D.R.: [1] Boundary Correspondence for Extremal Quasiconformal Mappings of the n-gon. PhD Dissertation, Univ. of Minn., June 1981.

Hubbard, J.H. and Masur, H.: [1] On the existence and uniqueness of Strebel differentials. Bulletin of the AMS 82, No. 1 (1976) 77-79.
- [2] Quadratic differentials and foliations. Acta Math. 142:3-4 (1979) 221-274.

Jenkins, J.A.: [1] Positive quadratic differentials in triply-connected domains. Ann. of Math. 53 (1951) 1-3.
- [2] On the existence of certain general extremal metrics. Ann. of Math. 66 (1957) 440-453.
- [3] Univalent functions and conformal mapping. Ergebnisse der Math. und ihrer Grenzgebiete, Neue Folge, Heft 18, 1-167, Springer Verlag (1958).
- [4] On the global structure of the trajectories of a positive quadratic differential. Illinois Journal of Math. Vol. 4, No. 3 (1960) 405-412.
- [5] A topological three pole theorem. Indiana Univ. Math. Journal, Vol. 21, No. 11 (1972) 1013-1018.
- [6] On quadratic differentials whose trajectory structure consists of ring domains. Proceedings of the 1976 Brockport symposium, Lecture notes in pure and appl. math. Vol. 36, 65-70. Edited by Sanford S. Miller.
- and Spencer, D.C.: [1] Hyperelliptic trajectories. Ann. of Math. Vol. 53, No. 1 (1951) 4-35.
- and Suita, N.: [1] On analytic selfmappings of Riemann surfaces II. Math. Ann. 209 (1974) 109-115.
- [2] On the representation and compactification of Riemann surfaces. Bull. Inst. of Math. Acad. Sinica 6, No. 2, Part II (1978) 423-427.

Jensen, G.: [1] Quadratic differentials. Chapter 8 in Univalent Functions by Ch. Pommerenke, Vandenhoeck & Ruprecht, Göttingen (1975) 205-260.

Kaplan, W,: [1] On the Three Pole Theorem. Math. Nachr. 75 (1976) 299-309.
- [2] On the trajectories of a quadratic differential, I. Comment. Math. Helv. 53 (1978) 57-72.
- [3] On the trajectories of a quadratic differential, II. Math. Nachr. 93 (1979) 259-278.

Keane, M.: [1] Interval Exchange Transformations. Math. Zeitschr. 141 (1975) 25-31.

Kerckhoff, Steven P.: [1] The Asymptotic Geometry of Teichmüller space. Topology 19, No. 1 (1980) 23-41.

Keynes, H.B. and Newton, D.: [1] A "minimal", non-uniquely ergodic interval exchange transformation. Math. Zeitschr. 148 (1976) 101-105.

Lehto, O.: [1] Riemann surfaces. Lecture Notes, University of Minnesota (1970) 1–113.
– and Virtanen, K.: [1] Quasiconformal mappings in the plane. Springer Verlag (1973)
Levine, H.I.: [1] Homotopic curves on surfaces. Proceedings AMS 14 (1963) 986–990.
Marden, A., Richards, I. and Rodin, B.: [1] On the regions bounded by homotopic curves. Pacific Journal of Math. 16, No. 2 (1966) 337–339.
– [2] Analytic selfmappings of Riemann surfaces. Journal d'Analyse Math. XVIII (1967) 197–225.
Marden, A. and Strebel, K.: [1] The heights theorem for quadratic differentials on Riemann surfaces. To appear in Acta Math.
– [2] On the ends of trajectories. To appear.
Masur, H.: [1] Jenkins-Strebel differentials with one cylinder are dense. Comment. Math. Helv. 54 (1979) 179–184.
– [2] Uniquely ergodic quadratic differentials. Comment. Math. Helv. 55 (1980) 255–266.
– [3] Interval Exchange Transformations and Measured Foliations. Preprint (1980) 1–55.
Minda, C.D. and Rodin, B.: [1] Extremal length, extremal regions and quadratic differentials. Comment. Math. Helv. 50, Fasc. 4 (1975) 455–475.
Nevanlinna, R.: [1] Uniformisierung. Grundlehren der Math. Wiss. Band LXIV, Springer Verlag (1953) zweite Aufl. (1967) 1–391.
Ohtsuka, M.: [1] On the behaviour of an analytic function about an isolated boundary point. Nagoya Math. J. 4 (1952) 103–108.
Pfluger, A.: [1] Lectures on conformal mapping. Indiana University (1966–67) 1–368.
Pommerenke, Chr.: [1] Univalent Functions. Vandenhoeck & Ruprecht, Göttingen und Zürich (1975) 1–376.
Reich, E.: [1] Sharpened distortion theorems for quasiconformal mappings. Notices of the AMS 10 (1963) 81.
Renelt, H.: [1] Konstruktion gewisser quadratischer Differentiale mit Hilfe von Dirichletintegralen. Math. Nachr. 73 (1976) 125–142.
Schaeffer, A.C. and Spencer, D.C.: [1] Coefficient regions for schlicht functions. AMS Colloquium Publications XXXV (1950) 1–311.
Schiffer, M.: [1] A method of variation within the family of simple functions. Proc. London Math. Soc. 44 (1938) 432–449.
Stoïlow, S.: [1] Leçons sur les principes topologiques de la théorie des fonctions analytiques. Gauthiers-Villars, Paris (1956) 1–194
Strebel, K.: [1] Die extremale Distanz zweier Enden einer Riemannschen Fläche. Ann. Acad. sci. Fen. 179 (1955) 1–21.
– [2] Zur Frage der Eindeutigkeit extremaler quasikonformer Abbildungen des Einheitskreises II. Comment. Math. Helv. 39 (1964) 77–89.
– [3] A uniqueness theorem for extremal quasiconformal mappings with fixed points and free boundary intervals. Proc. of the Conference on quasiconformal mappings, moduli and discontinuous groups, Tulane University (1965) 144–156.
– [4] Über quadratische Differentiale mit geschlossenen Trajektorien und extremale quasikonforme Abbildungen. Festband zum 70. Geburtstag von Rolf Nevanlinna (1965/1966) Springer Verlag, 105–127.
– [5] On quadratic differentials with closed trajectories and second order poles. Journal d'Analyse Math. XIX, Festband für L.V. Ahlfors (1967) 373–382.
– [6] Bemerkungen über quadratische Differentiale mit geschlossenen Trajektorien. Ann. Acad. Sci. Fenn. A.I. 405 (1967) 1–12.
– [7] On quadratic differentials and extremal quasiconformal mappings. Lecture Notes, University of Minnesota, Minneapolis (1967) 1–112.
– [8] Quadratische Differentiale mit divergierenden Trajektorien. Colloquium on mathematical analysis, Jyväskylä (1970). Springer Lecture Notes 419 (1973) 352–369.
– [9] On the trajectory structure of quadratic differentials. Discontinuous groups and Riemann surfaces, ed. by L. Greenberg, Ann. of Math. Studies 79 (1974) 419–438.
– [10] On the geometry of the metric induced by a quadratic differential I and II. Bull. Soc. Sci. Lettres Łódź 25, No. 2–3 (1975).
– [11] On quadratic differentials with closed trajectories on open Riemann surfaces. Commentationes in honorem Rolf Nevanlinna LXXX annos nato 1975, Ann. Acad. Sci. Fenn. A.I. 2 (1976) 533–551.

- [12] On the density of quadratic differentials with closed trajectories. Proceedings of the Rolf Nevanlinna Symposium on Complex Analysis, Silivri 1976, Istanbul (1978) 89–101.
- [13] On quasiconformal mappings of open Riemann surfaces. Comment. Math. Helv. 53 (1978) 301–321.
- [14] On the metric $|f(z)|^{\lambda}|dz|$ with holomorphic f. Complex Analysis Semester, Warsaw 1979. Banach Center Publ. 11 (1982) 323–336.
- [15] Vorlesungen über Riemannsche Flächen, Studia Mathematica, Skript 5, Vandenhoeck & Ruprecht, Zürich (1980) 1–120.
- [16] Quadratic differentials: A survey. To appear.

Teichmüller, O.: [1] Untersuchungen über konforme und quasikonforme Abbildungen. Deutsche Math. 3 (1938) 621–678.
- [2] Ungleichungen zwischen den Koeffizienten schlichter Funktionen. Sitzungsber. Preuss. Akad. Wiss., phys.-math. Kl. (1938) 363–375.
- [3] Extremale quasikonforme Abbildungen und quadratische Differentiale. Abh. Preuss. Akad. Wiss., math.-naturw. Kl. 22 (1939) 1–197.

Wiener, J.C.: [1] Isolated singularities of quadratic differentials arising from a module problem. Proceedings AMS 55 (1976) 47–52.

Subject Index

Ergebnisse der Mathematik und ihrer Grenzgebiete, 3. Folge

A Series of Modern Surveys in Mathematics

Band 1

A. Fröhlich

Galois Module Structure of Algebraic Integers

1983. X, 262 pages
ISBN 3-540-11920-5

Contents: Introduction. - Notation and Conventions. - Survey of Results. - Classgroups and Determinants. - Resolvents, Galois Gauss Sums, Root Numbers, Conductors. - Congruences and Logarithmic Values. - Root Number Values. - Relative Structure. - Appendix. - Literature List. - List of Theorems. - Some Further Notation. - Index.

Band 2

W. Fulton

Intersection Theory

1984. XI, 470 pages
ISBN 3-540-12176-5

Contents: Introduction. - Rational Equivalence. - Divisors. - Vector Bundles and Chern Classes. - Cones and Segre Classes. - Deformation to the Normal Cone. - Intersection Products. - Intersection Multiplicities. - Intersections on Non-singular Varieties. - Excess and Residual Intersections. - Families of Algebraic Cycles. - Dynamic Intersections. - Positivity. - Rationality. - Degeneracy Loci and Grassmannians. - Riemann-Roch for Non-singular Varieties. - Correspondences. - Bivariant Intersection Theory. - Riemann-Roch for Singular Varieties. - Algebraic, Nomological and Numerical Equivalence. - Generalizations. - Appendix A: Algebra. - Appendix B: Algebraic Geometry (Glossary). - Bibliography. - Notation. - Index.

Springer-Verlag
Berlin
Heidelberg
New York
Tokyo

Ergebnisse der Mathematik und ihrer Grenzgebiete, 3. Folge

A Series of Modern Surveys in Mathematics

Band 3

J. C. Jantzen

Einhüllende Algebren halbeinfacher Lie-Algebren

1983. V, 298 Seiten
ISBN 3-540-12178-1

Inhaltsübersicht: Einleitung. – Einhüllende Algebren. – Halbeinfache Lie-Algebren. – Zentralisatoren in Einhüllenden halbeinfacher Lie-Algebren. – Moduln mit einem höchsten Gewicht. – Annullatoren einfacher Moduln mit einem höchsten Gewicht. – Harish-Chandra-Moduln. – Primitive Ideale und Harish-Chandra-Moduln. – Gel'fand-Kirillov-Dimension und Multiplizität. – Die Multipliziztät von Moduln in der Kategorie $\mathcal{O}$. – Gel'fand-Kirillov-Dimension von Harish-Chandra-Moduln. – Lokalisierungen von Harish-Chandra-Moduln. – Goldie-Rang und Kostants-Problem. – Schiefpolynomringe und der Übergang zu den $\mathfrak{m}$-Invarianten. – Goldie-Rang-Polynome und Darstellungen der Weylgruppe. – Induzierte Ideale und eine Vermutung von Gel'fand und Kirillov. – Kazhdan-Lusztig-Polynome und spezielle Darstellungen der Weylgruppe. – Assoziierte Varietäten. – Literatur. – Verzeichnis der Notationen. – Sachregister.

Band 4

W. Barth, C. A. M. Peters, A. Van de Ven

Compact Complex Surfaces

1984. Approx. 10 figures. Approx. 320 pages
ISBN 3-540-12172-2

Contents: Introduction. – Standard Notations. – Preliminaries. – Curves on Surfaces. – Mappings of Surfaces. – Some General Properties of Surfaces. – Examples. – The Enriques–Kodaira Classification. – Surfaces of General Type. – $K3$-Surfaces and Enriques Surfaces. – Bibliography. – Subject Index.

Springer-Verlag
Berlin
Heidelberg
New York
Tokyo